MATHS
NOW!
2

RED ORBIT

MATHS NOW!
National Writing Group

JOHN MURRAY

Acknowledgements

The authors and publisher would like to thank all the teachers, schools and advisers who evaluated *Maths Now!* and whose comments contributed so much to this final version.

Particular thanks go to: David Ross, Mathematics Adviser, Wiltshire; John D Collins, Education Consultant and Inspector of Schools; Patrick Gallagher, Convent of Jesus and Mary RC High School, London; David J McLaren, Consultant in Mathematical Education; D J Bradwell, Desborough Comprehensive School, Maidenhead; David Bullock, Frodsham High School, Warrington; David Jolly, Antrim High School, Antrim; Maggie McCann, Kibworth High School, Leicester; Peter Marks, Ilfracombe College, Devon; Kevin Pankhurst, Pilton Community College, Barnstaple; Dr Philip Wanless, Belmont Comprehensive School, Durham.

Photo acknowledgements

Cover: Mark Harwood/Tony Stone Images; **p.10** Hank Morgan/Science Photo Library; **p.16** John Townson/Creation; **p.35** Mary Evans Picture Library; **pp.61, 93** John Townson/Creation; **p.108** reproduced from the 1998 1:50,000 Ordnance Survey map of Stirling & The Trossachs by permission of the Controller of HMSO © Crown Copyright; **pp.112, 113** John Townson/Creation; **pp.127, 128** Sally and Richard Greenhill; **p.141** © Michelin, from Map 60, 37th edition, 1998, Authorization no. 9901024; **p.142** © Michelin, from Map 916, 26th edition, 1998, Authorization no. 9901024; **p.149** © Michelin, from Map 970, 9th edition, 1998, Authorization no. 9901024; **p.155** © British Museum; **pp.156, 214, 225, 290, 291, 322, 342** John Townson/Creation; **p.349** Mary Evans Picture Library; **p.350** John Townson/Creation.

First published in 1999
by John Murray (Publishers) Ltd
50 Albemarle Street
London W1X 4BD

Layouts by Stephen Rowling/unQualified Design
Artwork by Oxford Illustrators Ltd
Cover design by John Townson/Creation

Typeset in 12/14pt Times by Wearset, Boldon, Tyne and Wear
Printed and bound by G. Canale, Italy

A CIP catalogue record for this book is available from the British Library.

ISBN 0 7195 7097 2
Teacher's File 2 Red Orbit 0 7195 7098 0

Contents

Introduction

By now you have already spent one year in secondary school. Do you remember all the maths you did last year? You are bound to have forgotten some of it, but don't worry. There is a lot of revision in this book – when a new topic is brought in, there is revision to connect it with what you did before.

In this book you will learn new things as well as revising the old! There are plenty of exercises to practise on, so that you become expert in the techniques, and there are also exercises which relate the mathematics you are doing to other areas of your life. If you keep your eyes open, you will see that there is mathematics all around you!

Symbols

Book 1

Use a calculator

Do not use a calculator

This is a particularly challenging question

Ma1

1 Ratio

This is a recipe for coconut macaroons, which you are going to make for a party.

Coconut macaroons

3 egg whites
pinch of salt
250 grams sugar
2 tablespoons flour
drop of vanilla essence
250 grams desiccated coconut

These amounts will make 20 macaroons. The day before the party, you hear that twice as many people will be coming, so you decide to make twice as many macaroons. It is easy to adjust the recipe. As there are going to be double the number of people, double the amounts of each ingredient.

6 egg whites
2 pinches of salt
500 grams sugar
4 tablespoons flour
2 drops of vanilla essence
500 grams desiccated coconut

A recipe tells you the ingredients you need for a dish. You can adjust the quantity you make, by multiplying or dividing the amounts of each ingredient by the same number.
 You can have 'recipes' for other things beside food. When you mix concrete, you must get the correct amount of each ingredient. A recipe for concrete might include:

1 bucket of cement
2 buckets of sand
3 buckets of gravel

If you want to make more concrete, just multiply each of the amounts by the same number. To make three times as much concrete, you need:

3 buckets of cement
6 buckets of sand
9 buckets of gravel

Exercise 1.1

Caldo Verde
Mushrooms with cumin & asafoetida
Terrine de Porc
Apricotina
Baklava

1 This recipe will make six servings of *Caldo verde*. Write down the ingredients for twelve servings.

> 4 medium potatoes
> 1 medium onion
> 100 grams kale leaves
> 12 cloves garlic
> 1 teaspoon salt
> 1 tablespoon olive oil

2 These are the ingredients for four servings of Mushrooms with cumin and asafoetida. Write down the ingredients for twelve servings.

> 800 grams mushrooms
> 2 tablespoons oil
> pinch of asafoetida
> $\frac{1}{2}$ teaspoon cumin
> 300 ml tomato sauce

3 Write down the ingredients for two servings of the recipe of question 2.
4 A kilogram of a mixture of birdseed contains 600 grams of millet and 400 grams of sunflower seed. How much sunflower seed do you need for two kilograms of the mixture? How much millet do you need for a third of a kilogram of the mixture?
5 A recipe for grapefruit marmalade uses two grapefruit for every lemon.
  a How many grapefruit do you need for four lemons?
  b How many lemons do you need for six grapefruit?
6 Jim runs a sports shop. For every three golf balls he sells, he gives away two golf tees. If Anne buys six golf balls, how many tees does she receive?
7 In Mindy's restaurant, they sell five cheesecakes for every three bagels.
  a How many cheesecakes do they sell if they sell 12 bagels?
  b How many bagels do they sell if they sell 15 cheesecakes?
8 Lucia is sharing beads with her little sister Fiona. Lucia's rule is 'two to you and three to me'. How many beads does Fiona get if Lucia gets 18 beads?

Suppose you want to make a particular dish, but have only a limited amount of one ingredient. Then you have to limit the other quantities accordingly.

Example The ingredients for Apricotina are:

> 300 grams dried apricots
> 160 grams butter
> 60 grams sugar
> 4 eggs

When you look in the fridge, you find that only 3 eggs are left. You must reduce the amounts of the other ingredients. You have 3 eggs instead of 4, so you have only $\frac{3}{4}$ of the number of eggs. You multiply the other amounts by $\frac{3}{4}$.

$$300 \times \tfrac{3}{4} = 225 \quad 160 \times \tfrac{3}{4} = 120 \quad 60 \times \tfrac{3}{4} = 45$$

So the reduced list of ingredients is:

> 225 grams dried apricots
> 120 grams butter
> 45 grams sugar
> 3 eggs

Exercise 1.2

1 This recipe is for *Terrine de porc*. Write down the ingredients if only two chops are available.

> 4 pork chops
> 750 grams of potatoes
> 150 ml white wine
> 2 cloves of garlic
> 4 juniper berries
> 120 grams ham

2 In question 1 of exercise 1.1 there was a recipe for *Caldo verde*. Write down the ingredients if 5 medium potatoes are to be used.

3 A type of concrete uses one bucket of cement, three buckets of sand and five buckets of gravel. Write down the amounts of cement and sand to go with 15 buckets of gravel.

4 To make 15 litres of green paint, you mix 9 litres of blue paint with 6 litres of yellow paint.
  a How much yellow paint do you need to go with 6 litres of blue paint?
  b How much blue paint do you need to go with 2 litres of yellow paint?

5 To make a bowl of fruit salad, you need 3 apples, 4 oranges, a pineapple, 2 pears, a melon and 3 bananas.
  a Write down the ingredients to go with 2 oranges.
  b Write down the ingredients to go with 3 pears.

6 Here is a list of ingredients to make Baklava for 20 portions.

> 240 g butter
> 180 g sugar
> 300 ml water
> 320 g chopped walnuts
> 320 g chopped almonds
> $\frac{1}{2}$ teaspoon ground cinnamon
> 500 g filo pastry

  a Write down the ingredients for 10 portions.
  b Danny has only 80 g of chopped walnuts and 200 g of filo pastry, but plenty of the other ingredients. How many portions can he make?
  c Danny rings his cousin Nick, who brings 80 g of chopped walnuts but no filo pastry. How many portions can he make now?

Constant ratio

In a recipe, the amounts of the ingredients may vary, but the **proportion** of each ingredient is constant. You can make a small amount of pastry for a single pie, or a lot of pastry for several pies. So the amounts of flour and fat vary greatly, from part of a kilogram to hundreds of kilograms.

But if it is the same sort of pastry, then the amounts of flour and fat vary at the same rate.

If you halve the amount of flour, you must also halve the amount of fat. If you treble the amount of fat, you must also treble the amount of flour. The amount of flour divided by the amount of fat is constant.

$$\frac{\text{amount of flour}}{\text{amount of fat}} \text{ is constant}$$

We say that the **ratio** between the amounts is constant. The ratio doesn't tell us how much of an ingredient there is; it tells us how much of it there is relative to the other ingredient.

Instead of using a fraction to write a ratio, we use a colon (two dots :) between the numbers. Suppose there is 3 kg of flour to 2 kg of fat. The ratio of flour to fat is

3:2

Note that the ratio does not have units. It is a ratio of pure numbers, which holds for all units. The units could be metric or Imperial. These amounts are all in the ratio 3:2.

3 ounces of flour to 2 ounces of fat
3 grams of flour to 2 grams of fat
3 tonnes of flour to 2 tonnes of fat

We can increase the amounts, provided that they increase at the same rate. We can multiply the amounts by the same number, and they will still be in the ratio 3:2. We can also divide the amounts by the same number without changing the ratio.

These amounts are still in the same ratio, 3:2.

30 ounces of flour to 20 ounces of fat   (multiplying by 10)
6 ounces of flour to 4 ounces of fat   (multiplying by 2)
9 ounces of flour to 6 ounces of fat   (multiplying by 3)
$1\frac{1}{2}$ ounces of flour to 1 ounce of fat   (dividing by 2)

Exercise 1.3

1 A type of brass is made from copper and zinc in the ratio 8:3. Copy and complete the following so that they are all in the ratio 8:3.
   a ___ grams of copper to 3 grams of zinc
   b 8 ounces of copper to ___ ounces of zinc

2 A packet of birdseed contains 4 kg of millet and 1 kg sunflower seed. Copy and complete the following so that they are in the same ratio.
   a 4 pounds of millet to ___ pound of sunflower seed
   b ___ grams of millet to 1 gram of sunflower seed

3 The brass of question 1 was made from copper and zinc in the ratio 8:3. Copy and complete the following so that they are still in the ratio 8:3.
   a 16 grams of copper to ___ grams of zinc   (multiplying by 2)
   b ___ grams of copper to 30 grams of zinc   (multiplying by 10)
   c ___ kg of copper to $\frac{3}{4}$ kg of zinc   (dividing by 4)
   d 24 ounces of copper to ___ ounces of zinc   (   )
   e ___ grams of copper to $\frac{3}{8}$ gram of zinc   (   )

4 The birdseed of question 2 was made from millet and sunflower seed in the ratio 4:1. Copy and complete the following so that they are in the same ratio.
   a 8 kg of millet to ___ kg of sunflower seed
   b 6 kg of millet to ___ kg of sunflower seed
   c ___ kg of millet to 2 kg of sunflower seed
   d ___ kg of millet to $\frac{1}{4}$ kg of sunflower seed

Finding the ratio

In mathematics, we always like to make things as simple as possible. So it is sensible to give a ratio in its simplest form. When you are given two whole number amounts, simplify them by dividing both by any common factor.

Suppose a pastry recipe calls for 150 grams of flour for 120 grams of fat. The ratio is

$$150:120$$

Both terms are multiples of 30. So if we divide them by 30, the ratio is still the same. It becomes

$$5:4$$

The numbers 5 and 4 have no common factor, so we have made the ratio as simple as possible.

You may have heard of the **gear ratio** of a car or bicycle. When two gear wheels meet, as in the diagram below, the gear ratio is the ratio between the numbers of teeth on the wheels.

Example The diagram below shows two gear wheels, which have 75 teeth and 50 teeth, respectively. Find the gear ratio, in its simplest form.

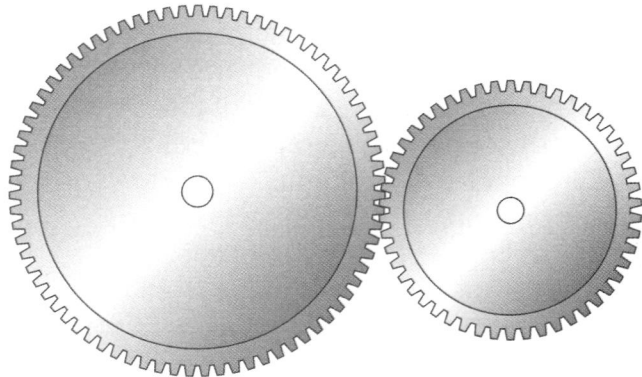

The ratio is $50:75$. Both of these can be divided by 25, giving $2:3$.
The gear ratio is $2:3$.

Exercise 1.4

1 Simplify the following ratios as far as possible.
   a 4:6   **b** 14:21   **c** 150:25
2 A recipe for a Manhattan cocktail mixes 160 ml of rye whiskey with 80 ml of sweet vermouth. What is the ratio of whiskey to vermouth, in its simplest form?
3 An ingot of bronze contains 20 kg of copper and 5 kg of tin. What is the ratio of copper to tin, in its simplest form?
4 Two gear wheels have 80 and 60 teeth, respectively. Find the gear ratio in its simplest form.
5 A necklace has 35 brown beads and 28 green beads. What is the ratio of the number of brown beads to the number of green beads, in its simplest form?
6 Three quarters of my garden is made up of lawn, the rest being used for vegetables. What is the ratio of lawn to vegetables?
7 A school has 1728 pupils and 96 teachers. Find the 'pupil–teacher' ratio, in its simplest form.
8 A junior golf club has 140 members, of whom 80 are boys. Find the ratio of boys to girls, in its simplest form.

The ratio 1 : *n*

Some ratios are written as 1:n, where n is a whole number. This means that one quantity is n times the other. So, given one quantity, you can find the other quantity, either by dividing by n or by multiplying by n.

Examples   The teacher–pupil ratio of a school is 1:20. If there are 25 teachers, how many pupils are there?

Think carefully about whether to multiply by *n* or divide by *n*.

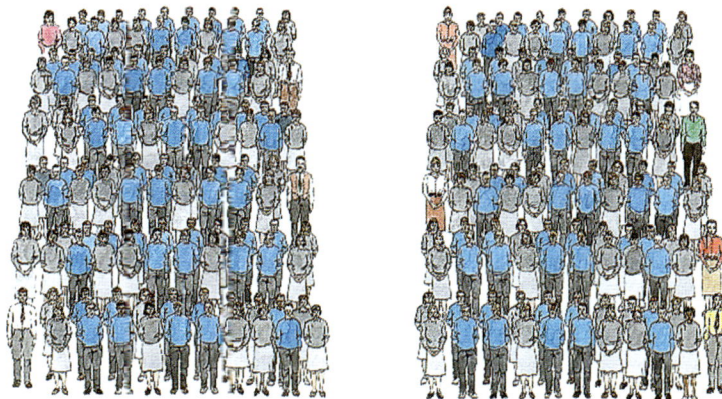

Multiply 25 by 20. 25 × 20 = 500.
   There are 500 pupils.

> 1 teacher for 12 pupils is the same as 12 pupils for 1 teacher.

Another school has a teacher–pupil ratio of $1:12$. If there are 840 pupils, how many teachers are there?

Divide 840 by 12. $840 \div 12 = 70$.
    There are 70 teachers.

Exercise 1.5

1   A type of steel contains carbon and iron in the ratio $1:40$. How much iron should go with 3 kg of carbon?

2   For the steel of question 1, how much carbon should go with 200 kg of iron?

3   In an army, the ratio of officers to other ranks is $1:10$. If there are 2500 officers, how many are there in the other ranks?

4   A law states that the ratio of adults to children in a playgroup should be at least $1:4$. If there are 5 adults, how many children could they look after?

5   For the playgroup of question 4, what is the least number of adults needed to be in charge of 28 children?

6   For the playgroup of question 4, what is the least number of adults needed to be in charge of 30 children?

7   In a batch of eggs, the ratio of cracked eggs to uncracked eggs is $1:16$. If there are 5 cracked eggs, how many uncracked eggs are there?

8   For the batch of eggs of question 7, how many cracked eggs are there if there are 320 uncracked eggs?

9   Geoff is going on a sponsored run. When he asks people to sponsor him, he finds that the ratio of people who will sponsor him to those who won't is $1:3$. If 40 people sponsor him, how many won't sponsor him?

10   In question 9, how many people must Geoff ask to find 50 sponsors?

Scale diagrams

A ratio of $1:n$ is often used when we draw something. An accurate picture of a building or a room is a **scale diagram**. The scale is the ratio of lengths in the diagram to lengths in the real building. The units (centimetres, metres, etc.) used for both lengths must be the same.

lengths in diagram : real lengths = scale

Examples   The diagram below shows a scale diagram of a room. The diagram is 3 cm by 4 cm, and the scale is $1:200$. What size is the real room?

door

window

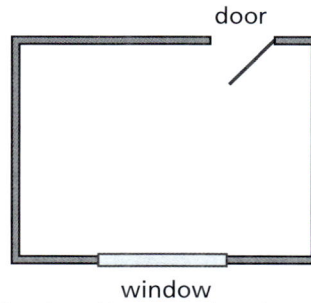

Multiply the lengths in the diagram by the scale.

$$200 \times 3\,cm = 600\,cm = 6\,m \qquad 200 \times 4\,cm = 800\,cm = 8\,m$$

The real room is 6 m by 8 m.

....

A scale diagram of a field is at a scale of $1:500$. If the field is 200 m long, find the length of the scale diagram.

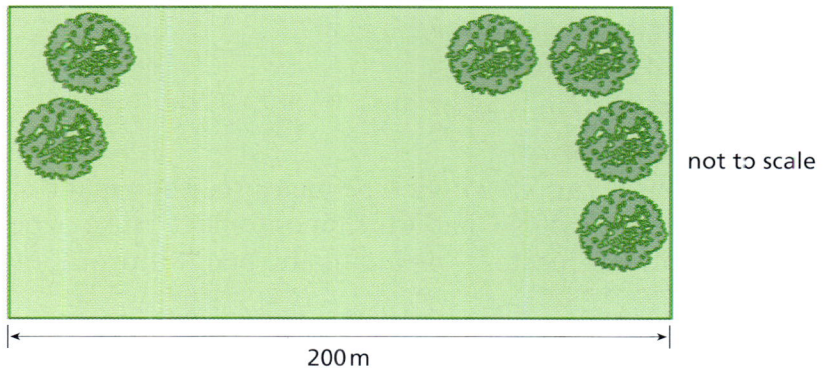

not to scale

200 m

Going from the real field to the scale diagram, we *divide* by 500.

$$200\,m \div 500 = 0.4\,m = 40\,cm$$

The length of the scale diagram is 40 cm.

....

A room is 6 m long. In a diagram of the room, its length is 50 cm. What is the scale of the diagram?

Remember:

The units in ratios must be the same.

Convert 6 m to 600 cm.
The scale is $50:600$. Divide by 50, to obtain $1:12$.
The scale is $1:12$.

Exercise 1.6

1 A scale diagram of a room is at a scale of $1:80$. If the length of the room in the diagram is 12 cm long, find the length of the real room.
2 A scale diagram of a building plot is in the ratio $1:400$. If the plot is 1.2 m wide on the diagram, what is the width of the real plot?

3 A model of a car is at a scale of 1:100. If the model is 4 cm long, what is the length of the real car?

4 A toy soldier is at a scale of 1:90. If the toy is 2 cm high, what is the height of the real soldier?

5 A scale diagram of a room is at a scale of 1:20. If the room is 400 cm wide, what is its width on the diagram?

6 A scale diagram of a house is at a scale of 1:40. The real height of the house is 10 m. What is its height on the diagram?

7 A model of an engine is at a scale of 1:150. The real engine is 3 m wide. How wide is the model?

8 The animals of a toy zoo are made in the ratio 1:60. The toy tiger is 6 cm long. How long is a real tiger?

9 A room is 400 cm long. The length of a scale diagram of the room is 50 cm. What is the scale of the diagram?

10 A building is going to be 28 m high. On the architect's diagram, the building is 2 m high. What is the scale of the plan?

Exercise 1.7 Ma 1

Measure your bedroom with a tape measure. Then make a scale diagram of the room. Choose a scale so that it fits on one sheet of A4 paper. You can make the diagram more elaborate by including the bed, wardrobe, etc.

Increasing ratio

Ratios are very useful when we want to express quantities that are increasing or decreasing. If a quantity is *increased* in the ratio 1:5, then its value is *multiplied* by 5.

old amount : new amount
1 : 5

There are many machines which enable us to lift heavy objects, like the ones in the photo. All these machines increase our force in a fixed ratio.

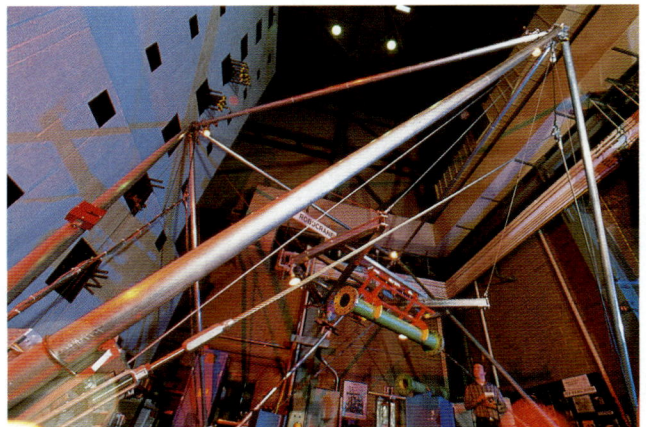

● *There is a constant ratio between the weight we lift and the force we apply. This ratio is the* **mechanical advantage***.*

Stonehenge is built of huge blocks of stone. The prehistoric Britons who built Stonehenge may have used levers to raise the stones.

Example   A lever is such that a weight of 20 kg can lift a weight of 300 kg. What is the mechanical advantage?

Using the lever, what can we lift with a weight of 30 kg?

The mechanical advantage is 300:20. Simplify this ratio by dividing both sides by 20.

The mechanical advantage is 15:1.

With a weight of 30 kg, we could lift 15 × 30 kg, that is, 450 kg.

Exercise 1.8

1 Increase 10 in the ratio 1:2.
2 Increase 15 in the ratio 1:3.
3 Increase 40 in the ratio 1:5.
4 Increase 18 in the ratio 1:6.
5 Increase 7 in the ratio 1:8.

6 An electricity transformer increases the current in the ratio 1:12. If the current is 2 amps before the increase, what is it after?

7 A system of pulleys has a mechanical advantage of 4:1. What weight can we lift with a weight of 50 kg?

8 A lever has a mechanical advantage of 16:1. What weight can we lift with a weight of 40 kg?

9 A system of pulleys can raise 200 kg with a weight of 25 kg. What is the mechanical advantage of the system?

10 A small person can stop a big car! The hydraulic brakes on a car can exert a braking force of 10 000 newtons, for a force on the brake pedal of 200 newtons. What is the mechanical advantage of the brakes?

Decreasing ratio

If a quantity is *decreased* in the ratio 5:1, then its value is *divided* by 5.

old amount : new amount
5 : 1

Exercise 1.9

1 Decrease 100 in the ratio 2:1.
2 Decrease 45 in the ratio 3:1.
3 Decrease 75 in the ratio 5:1.
4 Decrease 20 in the ratio 10:1.
5 Decrease 18 in the ratio 3:1.
6 A company decreases its debt in the ratio 4:1. If it used to owe £60 000, what does it owe now?
7 A transformer changes voltage in the ratio 80:1. What does it decrease a voltage of 20 000 volts to?
8 A system of pulleys has a mechanical advantage of 30:1. What weight is needed to lift 900 kg?

9 The brakes of a car have a mechanical advantage of 80:1. What force on the brake pedal is needed to brake the car with a force of 16000 newtons?

10 A computer chip is made smaller in the ratio 5:1. If it used to be 1.5 cm long, what length is it now?

Division in ratio

Let's return to our example of pastry which consists of flour and fat in the ratio 3 2. How much of each do you need to make 20 kg of pastry?

There are 3 parts of flour for every 2 parts of fat,

3 parts of flour : 2 parts of fat

making 5 parts in total. So there are 5 equal parts of pastry material.

If we want to make 20 kg in total, then each of the 5 parts must weigh 20 kg ÷ 5, that is, 4 kg. Each part is 4 kg.

There are 3 parts of flour, which must weigh 3 × 4 kg, that is, 12 kg.

There are 2 parts of flour, which must weigh 2 × 4 kg, that is, 8 kg.

So that is what we need, 12 kg of flour and 8 kg of fat.

We can check our answer.

The sum of 12 kg and 8 kg is 20 kg, the total amount of pastry.

The ratio of flour to fat is 12:8, which is the same as 3:2.

This is **division in ratio**.

We wanted to divide 20 in the ratio 3:2.

We added the terms of the ratio, getting 5.

We divided 20 by 5, obtaining 4.

The amounts were then found by multiplying 3 by 4 and then 2 by 4.

Example In a will, an aunt leaves money to her nephew and niece in the ratio 4:5. If the total amount she left was £27000, how much does each get?

Here there are 4 + 5, that is 9, equal parts. Divide 27000 by 9, giving 3000. So each part is worth £3000. The nephew gets 4 of these parts, that is, £12000. The niece gets 5 of the parts, that is, £15000.

Her nephew gets £12000 and her niece gets £15000.

Exercise 1.10

1 Divide 10 in the ratio 2:3.
2 Divide 12 in the ratio 1:3.
3 Divide 24 in the ratio 5:3.
4 Divide 18 in the ratio 4:5.
5 An alloy contains copper and tin in the ratio 11:3. How much of each metal is needed to make 70 kg of the alloy?
6 A vinaigrette mixes oil and vinegar in the ratio 4:1. How much of each do you need to make 30 cm³ of vinaigrette?
7 A school contains boys and girls in the ratio 5:7. If there are 360 pupils in total, how many are girls?
8 Two people share a flat. They pay the expenses in the ratio 6:7. How much should each pay towards a bill of £260?
9 The teacher–pupil ratio of a school is 1:17. How many teachers are there if there are 900 people in the school?
10 Two authors collaborate on the writing of a textbook. The first author writes 160 pages, and the second writes 80 pages. What is the ratio of their contributions, in its simplest form? The book earns £18 000 in royalties. If they share the royalties in the same ratio as the contributions, how much does each get?

SUMMARY

■ A **ratio** expresses the relative amounts of two quantities. A ratio of 2:3 means that 2 units of the first ingredient go with 3 units of the second ingredient.
■ If quantities are in the ratio 1:n, then the second quantity is n times the first quantity. To go from the first quantity to the second, multiply by n. To go from the second quantity to the first, divide by n.
■ In **scale diagrams** the scale is the ratio of lengths in the diagram to the real lengths. The lengths must be measured in the same units.
■ To increase a quantity in the ratio 1:n, multiply by n.
■ To decrease a quantity in the ratio n:1, divide by n.
■ To divide a quantity in a ratio, add the terms of the ratio to find the number of equal parts. Divide the quantity by this, then multiply by the terms of the ratio.

Exercise 1A

1 To make a Grecian cocktail, blend 2 measures peach juice, 1 measure orange juice and $\frac{1}{2}$ measure of lemon juice. Write down the measures of ingredients needed to make four cocktails.

2 Copy and complete the following so that they are all in the ratio 3:2.
  a ___ grams to 2 grams
  b 30 grams to ___ grams
  c ___ grams to 16 grams
  d 1 gram to ___ grams
3 Simplify the ratio 15:20 as far as possible.
4 A computer and printer cost £1200 and £800, respectively. Find the ratio of their costs, in the simplest form.
5 A squash club contains adults and children in the ratio 3:1. If there are 50 children, how many adults are there?
6 A college is funded by fees and by the government in the ratio 1:5. If the government provides £800000, how much comes from fees?
7 A scale diagram of a room is in the ratio 1:4. If the diagram is 80cm long, what is the length of the room?
8 The mechanical advantage of a pulley system is 40:1. What weight is needed to raise a weight of 1600kg?
9 Divide £25 in the ratio 2:3.
10 The population of a state is divided between town dwellers and country dwellers in the ratio 3:1. If the total population is 12000000, how many are town dwellers?

Exercise 1B

1 To make Royal icing, mix 400 grams of sugar, 2 egg whites and the juice of $\frac{1}{2}$ lemon. Write down the ingredients if 3 egg whites are used.
2 George and Helen contribute £1000 and £800, respectively, to an investment. The value of the investment increases to £3600. How much should each receive?
3 Simplify the ratio 200:95.
4 A farm is divided between wheat and pasture in the ratio 3:1. If there are 900 acres of wheat, how much pasture is there?
5 A transformer increases electric current in the ratio 1:4. If the input current is 3 amps, what is the output current?
6 What input current is needed for the transformer of question 5 to produce an output current of 80 amps?
7 A bridge is to be 500m wide. The engineer's scale drawing of the bridge is 2m wide. What is the scale of the drawing?
8 When rice is cooked in water, it increases its weight in the ratio 1:2. If you start with 200 grams of raw rice, what does it weigh when it is cooked?
9 There are 800 pupils in a school, consisting of Seniors and Juniors in the ratio 9:7. How many Seniors are there?
10 At the end of a course there are two exams, with available marks in the ratio 1:4. If there are 100 marks in total, how many are there from each exam?

Exercise 1C Ma1

Many of the examples of this chapter have involved mixtures of metals, called alloys. There are many different alloys: look up the composition of some of the following. Make a display poster of the ratio of the metals.

pewter, Britannia metal, German silver, pinchbeck, amalgam, brass, bronze, solder

Exercise 1D Ma1

Do you or a friend own a bike with several gears? In different numbered gears, how often does the rear wheel turn for each turn of the pedals? How is this connected with the number of teeth on the sprockets? Investigate, and write a report. You might find it helpful to fill in a table like the one below.

gear	turns of rear wheel	teeth on pedal sprocket	teeth on rear wheel sprocket
1			
2			
.			
.			
.			

2 Angles

Aidan and Pat are lost on a moor.

We have to get to Flimhurst; that's where we're staying.

They find a milestone telling them that Flimhurst is 5 miles away.

That's all right, that'll take us just over an hour.

FLIMHURST 5 MILES

No, it isn't all right – we know it's five miles away, but in which direction? Which way do we go?

Remember:

An angle measures the amount of turn. A full turn is 360°.
A half turn is 180° A quarter turn (a right-angle) is 90°.

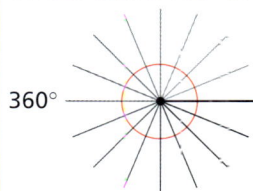

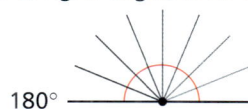

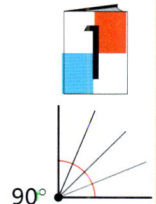

360°   180°   90°

There are special sorts of angle – do you remember their names?
Here is a quick revision exercise to see how much you remember.

Exercise 2.1

(Mainly revision)

1 Is the angle 43°:
  a acute   **b** obtuse   **c** reflex?
2 Is the angle 127°:
  a acute   **b** obtuse   **c** reflex?
3 Is the angle 225°:
  a acute   **b** obtuse   **c** reflex?

For the next four questions, copy and complete the statements, replacing the ___s by numbers.

4 An acute angle is less than _____°.

5 An obtuse angle is between _____° and _____°.

6 A reflex angle is greater than _____°.

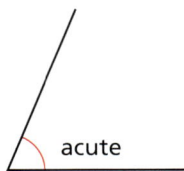

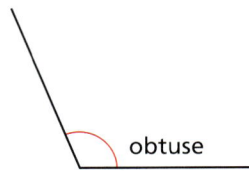

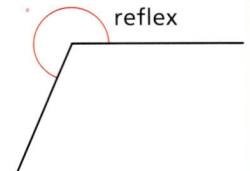

7 The angles along a straight line add up to _____°.

8 Is each of these angles *acute*, *obtuse* or *reflex*?

a   b   c

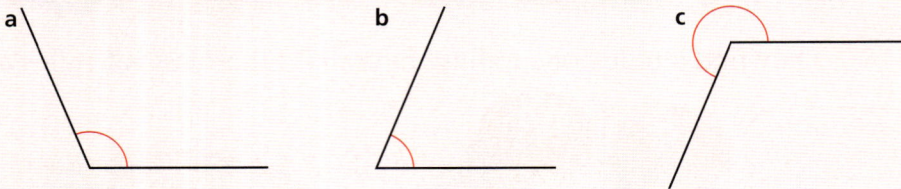

How confident were you in your answers? These are the definitions.

> **Remember:**
>
> *An acute angle is less than 90°.*
>
> *An obtuse angle is between 90° and 180°.*
>
> *A reflex angle is greater than 180°.*
>
> reflex
>
> acute   obtuse

Measuring angles

> **Remember:**
>
> *The outside set of numbers on a protractor measures clockwise angles and the inside set measures anti-clockwise angles.*

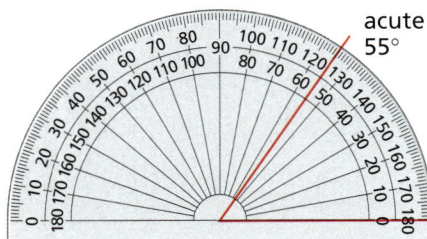

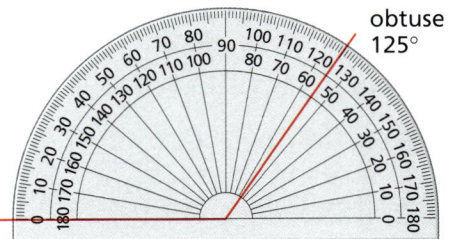

The instrument used to measure and draw angles is the **protractor**. At each point along the rim of the protractor are two angles, one acute and one obtuse. When you measure or draw an angle, be sure which type of angle it is. The diagram below shows how to measure acute and obtuse angles.

acute 55°   obtuse 125°

Suppose you want to measure a reflex angle. If you have a protractor which goes full circle, that is, from 0° up to 360°, then measuring a reflex angle is straightforward.

If you have a protractor which only goes up to 180°, then measure the acute or obtuse angle on the other side. Subtract this from 360°.

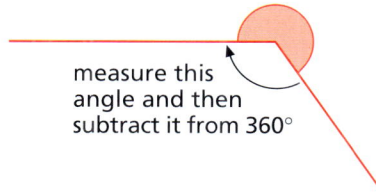

measure this angle and then subtract it from 360°

Example   Measure the reflex angle $\angle AXB$ below.

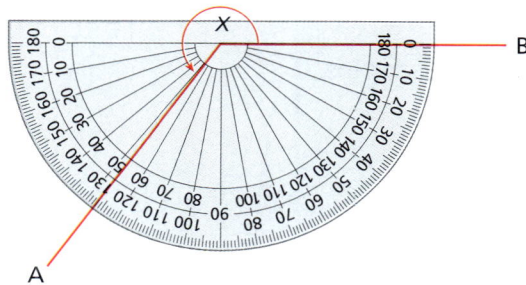

Use an ordinary protractor to find that the obtuse angle on the other side is 127°. Subtract from 360°.

$$360 - 127 = 233$$

So the reflex angle $\angle AXB$ is 233°.

To draw a reflex angle, subtract it from 360°. Draw this angle, and take the other side.

Example   Draw the angle 321°.

Subtract 321° from 360°, obtaining 39°. Draw this angle using an ordinary protractor. The reflex angle is the angle marked.

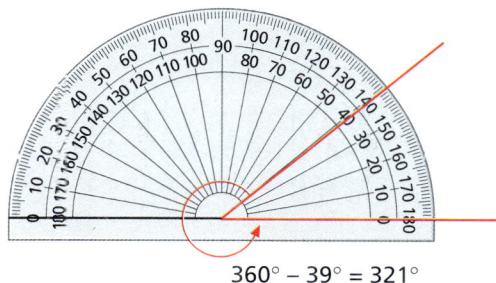

360° − 39° = 321°

Remind yourself of the use of the protractor by doing the next exercise.

Exercise 2.2

(Mainly revision)

1 Measure these angles.

   a 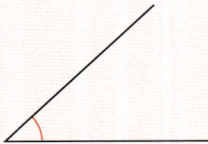   **b** 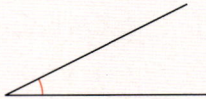   **c**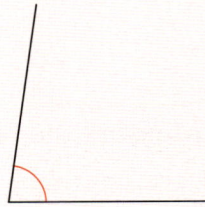

2 Measure these angles.

   a 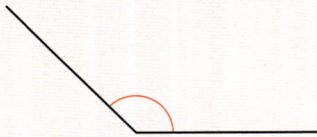   **b** 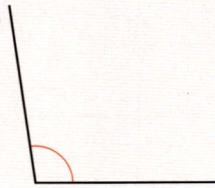   **c**

3 Measure these angles.

   a 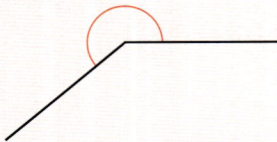   **b** 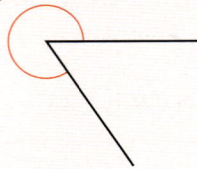   **c**

4 Measure these angles.

   a 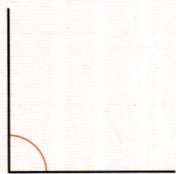   **b**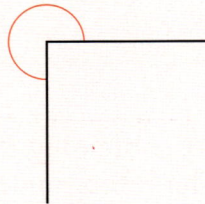

5 Draw these angles.
   a 38°   **b** 62°   **c** 88°

6 Draw these angles.
   a 132°   **b** 111°   **c** 98°

7 Draw these angles.
   a 280°   **b** 307°   **c** 355°

8 Draw these angles.
   a 190°   **b** 217°   **c** 258°

9 Draw these angles.
   a 90°   **b** 180°   **c** 270°

Equal angles

In Book 1, we found that certain pairs of angles are equal.
Opposite angles. When two straight lines cross, the opposite angles are equal. In the diagram

$a = b$

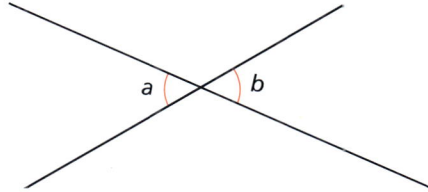

Look for X shapes.

A straight line crossing two parallel lines is a **transversal**.
Corresponding angles. The angles on the same side of the transversal and of the parallel lines are equal. In the diagram

$c = d$

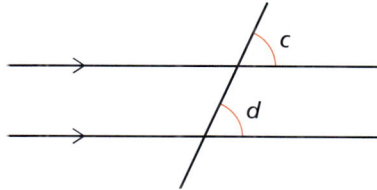

Look for F shapes.

Alternate angles. The angles on opposite sides of the transversal and of the parallel lines are equal. In the diagram

$e = f$

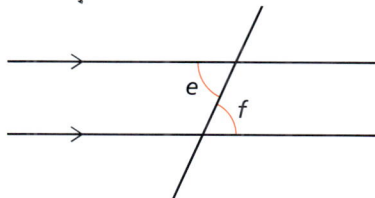

Look for Z shapes.

Can you see the letters X, F and Z in this diagram?

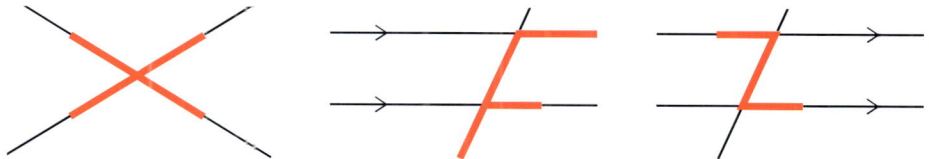

Exercise 2.3

(Mainly revision)

For questions 1 to 6, find the size of the angles marked with a letter. In each case say whether the angles are opposite, corresponding or alternate.

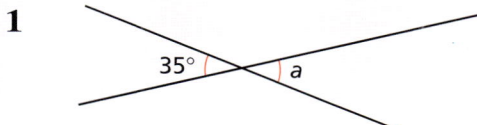

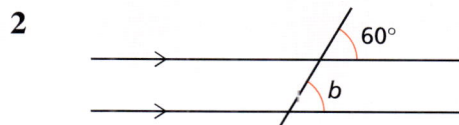

1

35° a

2

60° b

3

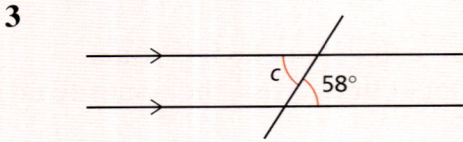

4

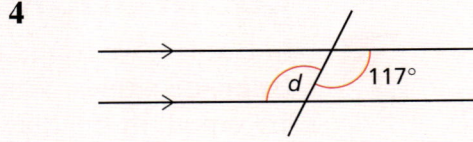

5

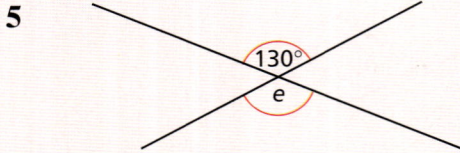

6

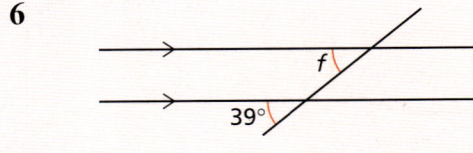

7 Write down the size of each angle marked with a letter in the diagram below.

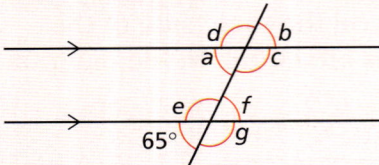

8 Write down the size of each angle marked with a letter in the diagram below.

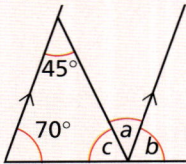

9 Calculate angle $2x$ in the diagram below.

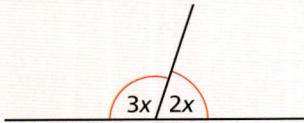

10 Calculate angle x in the diagram below.

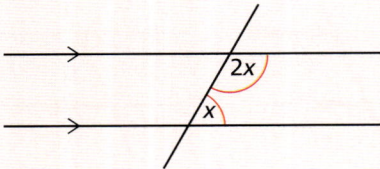

Bearings

When you are out at sea, and want to get to port, you need to know two things: how far it is to port, and the direction in which you should go. It isn't much help if you only know the distance – you could set off in completely the wrong direction!

The **bearing** tells you which way to go. Bearings are measured from North. Turn clockwise from North until you have the direction you want. The angle which you have turned through is the bearing. A bearing can be any angle up to 360°. For consistency we always write a bearing with three digits, for example 057° rather than 57°.

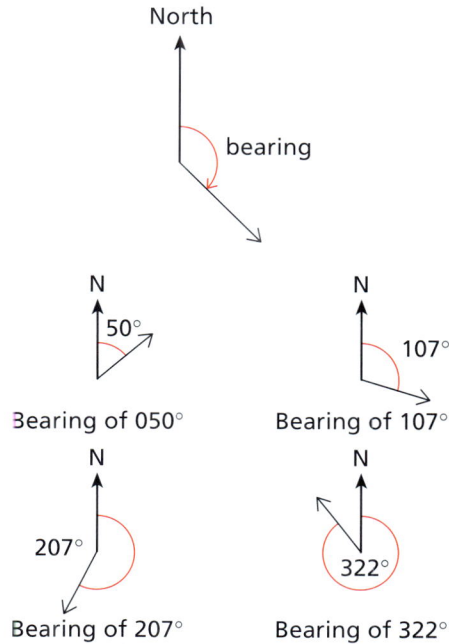

North

bearing

Bearing of 050°

Bearing of 107°

Bearing of 207°

Bearing of 322°

Notice that in the last example the angle goes round clockwise from North, to 322°. It doesn't go round anti-clockwise, to 038°. These directions are known as **three-figure bearings**.

Exercise 2.4

Using a protractor, find the bearings of these directions.

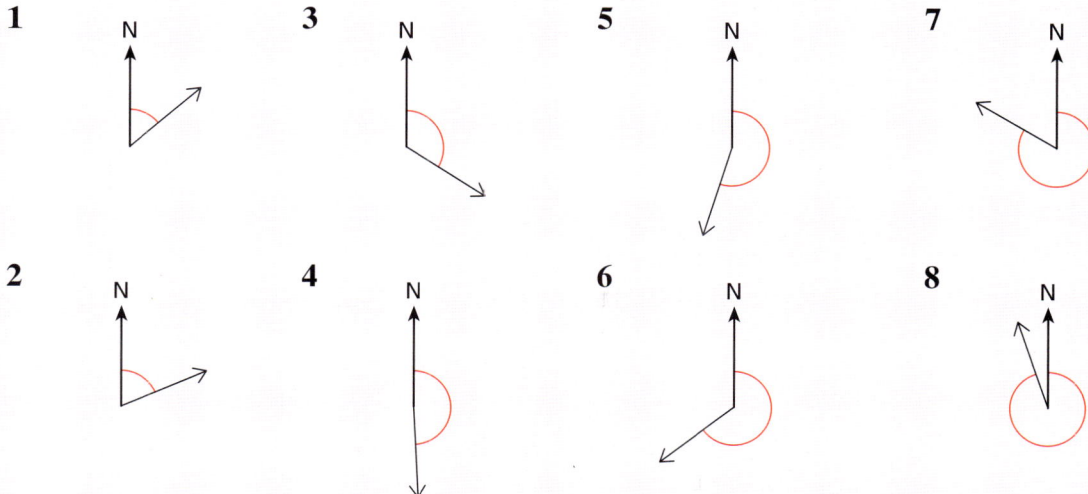

1

N

3

N

5

N

7

N

2

N

4

N

6

N

8

N

9 From a point *P*, a coastguard sees an aircraft at *A* on a bearing of 150°. At the same time he sees a boat at *B* on a bearing of 190°. Calculate the size of ∠*APB*.

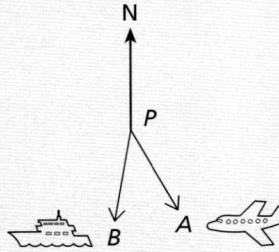

10 The diagram shows the position of an aircraft at *A*. Use your protractor to find the bearing on which the aircraft must fly to reach the airport at *P*.

Another way to find directions is by using the **points of the compass**. We usually show North as upwards. South is then downwards, East to the right and West to the left.

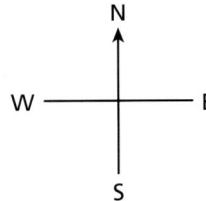

Half-way between North and East is North-East (NE). The other half-way points are SE, SW and NW.

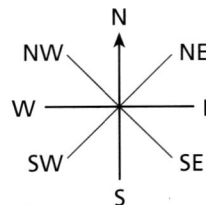

Half-way between North and North-East is North-North-East (NNE). The other points are ENE, ESE, SSE, SSW, WSW, WNW, NNW.

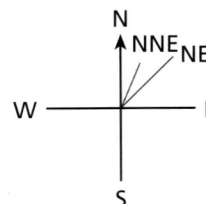

You can convert bearings into points of the compass.

North 000°   (or 360°)
East   090°
South 180°
West 270°

(Notice again that we measure bearings *clockwise* from North. West is at 270°, not 090°.)

North-East is half-way between North and East. So, to find its bearing, find half-way between 000° and 090°, which is 045°. For SE, SW and NW just add 45° to the bearings of E, S and W.

North-North-East is half-way between North and North-East. Half of 45 is 22½. So, to find the bearings of NNE, ENE, ESE, etc., just add 22½° to the bearings of N, NE, E, etc.

Exercise 2.5

1 Copy and complete the following conversion table.

point of the compass	bearing
North-East	045°
South-East	_____°
South-West	_____°
North-West	_____°

2 Copy and complete the following table.

point of the compass	bearing
NNE	022½°
ENE	_____°
ESE	_____°
SSE	_____°
SSW	_____°
WSW	_____°
WNW	_____°
NNW	_____°

3 Half-way between N and NNE is North by North-East (N by NE). What is the bearing of this direction?

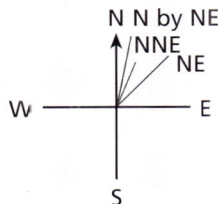

4 What are the angles between these directions?
  a N and E **b** NE and S **c** NE and SSW?
5 I am standing facing West. I turn 45° clockwise. What direction am I facing now?
6 A spinner rotates clockwise at a rate of 180° per second. It starts off pointing North. Which way is it pointing after
  a 2 seconds **b** $\frac{1}{2}$ second **c** $\frac{3}{4}$ second?

Setting out bearings

If you know the bearing you want to travel on, you can mark its direction on a map. Notice whether the bearing is acute, obtuse or reflex. Be especially careful with bearings greater than 180°, that is, reflex angle bearings. At the beginning of this chapter you had a lot of practice at drawing angles of all sizes.

Examples On the diagram below you are at point *A*. The North direction is shown. You want to travel on a bearing of 110°. Draw a line to show your journey.

N
↑
A •

The bearing is a bit more than 90°, so the direction will be a bit South of East. Place your protractor on the North line, with its centre at *A* and with the curve of the protractor to the right. Put a dot at 110° measured clockwise from North. Join *A* to this dot, and you have a direction on a bearing of 110°.

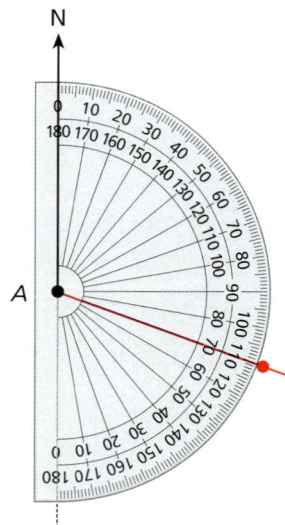

On this diagram you are at point *B*. You want to travel on a bearing of 326°. Draw your journey.

N

● *B*

The bearing is greater than 180°, so assume that you have already turned 180° and place your protractor with the curve to the left. Subtract 326 from 360, giving 34. Draw a dot at 34° measured anti-clockwise. Join *B* to this dot, and you have a direction on a bearing of 326°.

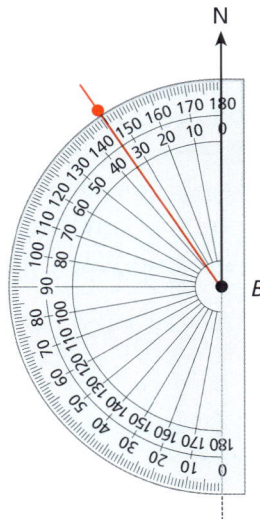

Exercise 2.6

In questions 1 to 4, you start at point *X* and move on the bearing given. In each case make a copy of the diagram and draw a line showing your journey.

N

1 A bearing of 048°.
2 A bearing of 123°.
3 A bearing of 229°.
4 A bearing of 354°.

● *X*

5 You are standing at the top *T* of a hill and facing North. What is the bearing of
  a the palm tree
  b the hut
  c the pit?

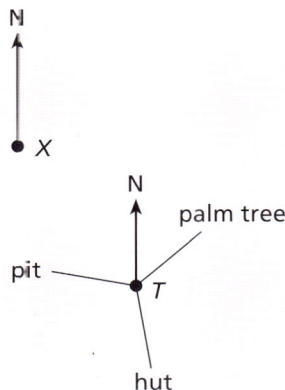

N
pit —
palm tree
● *T*
hut

6 From her house, Anna can see
the football pitch, the café, the
canal bridge and the pond.
Calculate the value of *x* and so
find the bearings of each of these
places from Anna's house.

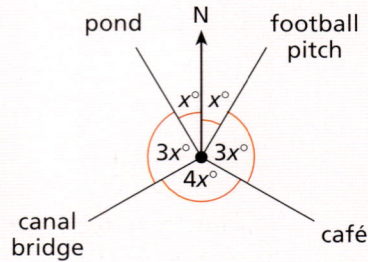

The bearing of an object doesn't tell you how far away it is. But if
you take the bearing of the object from one position, then walk
further on and take the bearing from another position, then you
can mark on a map where the object is.

Similarly, if you take the bearing of two objects from you, you
can find where you are!

To find the *bearing of A from B*, imagine that you are standing
at *B* and looking towards *A*. To find the *bearing of B from A*,
imagine you are standing at *A* and looking towards *B*. So you are
facing in the opposite direction.

*Be very careful
with bearings. Make
sure you are facing in
the right direction!*

*If you know the
bearing of A from B,
then the bearing of
B from A is the
back bearing.*

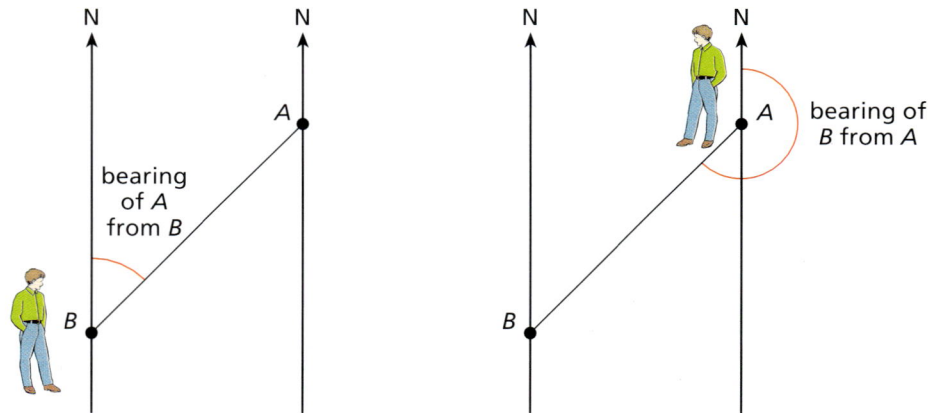

As you have turned through 180°, just add or subtract 180° to
find the other bearing.

If the bearing of *A* from *B* is 30°, the bearing of *B* from *A* is
30° + 180°, that is, 210°.
If the bearing of *A* from *B* is 200°, the bearing of *B* from *A*
is 200° − 180°, that is, 20°.

Examples A ship is out at sea. You take the bearing of the ship from two
positions, *A* and *B*. From *A*, the bearing is 040°. From *B*, the
bearing is 350°. Mark the position of the ship on the map.

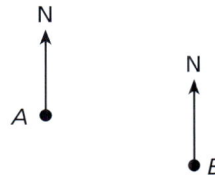

Use your protractor to set out a bearing of 040° from *A*. Set out a bearing of 350° from *B*. (Be careful with this, as it is a reflex angle.) Where the lines cross is the position of the ship, marked with *S* on the map below.

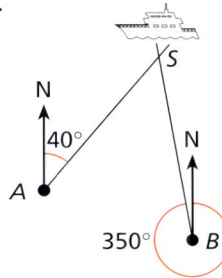

You are lost on a moor. But you can see two distant church spires which you recognise. The bearing of spire *A* is 110°, and the bearing of spire *B* is 155°. Mark on the map where you are.

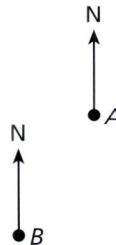

You have to do a bit of arithmetic here. If you are walking towards spire *A*, you are walking along a bearing of 110°. Suppose you turn round to walk in the opposite direction, that is, from *A* towards your present position. You will have turned through 180°, so you will be walking along a bearing of 110° + 180°, that is, 290°. So the bearing of your present position from *A* is 290°. Similarly the bearing of your present position from *B* is 155° + 180°, that is, 335°.

Mark out bearings of 290° from *A* and 335° from *B*, as shown below. You are at the crossing point of these lines, marked *Y*.

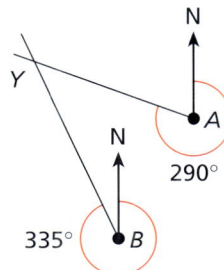

Exercise 2.7

1 The bearing of a tower is measured as 081° from point X, and as 010° from point Y. Make a copy of the map below and mark on it the position of the tower.

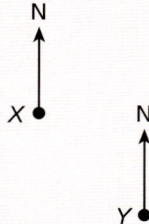

2 On an ancient manuscript you read that a treasure is buried at a place which is on a bearing of 320° from the church A, and on a bearing of 260° from the tower B. Make a copy of the map below, and mark the position of the treasure.

3 Rose is out hill walking. She takes the bearings of two hilltops and finds them to be 085° and 309°. What is the bearing of Rose from the hilltops?

4 You are out at sea. You see one lighthouse A on a bearing of 045°, and a second lighthouse B on a bearing of 146°. Make a copy of the map below and mark your position on it.

5 On the African plain you can see Mt Kilimanjaro on a bearing of 230° and Mt Kenya on a bearing of 309°. Make a copy of the map below and mark your position on it.

SUMMARY

- An angle gives the amount of turn between two lines. You can measure and draw angles with a protractor.
- When two straight lines cross, the **opposite angles** are equal.
- A **transversal** is a line crossing two parallel lines. The **corresponding angles**, on the same side of the transversal and of the parallel lines, are equal. The **alternate angles**, on either side of the transversal within the parallel lines, are equal.
- A **bearing** gives a direction. Bearings are measured clockwise from North. A bearing is always written with three digits.

Exercise 2A

1 Is the angle 163° acute, obtuse or reflex?

2 Use a protractor to measure this angle.

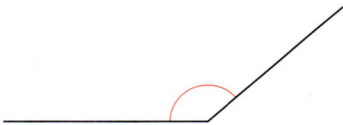

3 Calculate angle x.

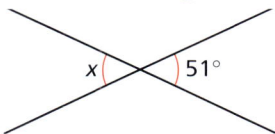

4 Calculate angle y.

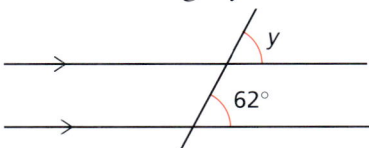

5 Calculate angle z, giving your reasoning.

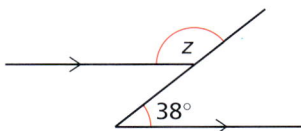

6 A and D are on opposite sides of BC. $\angle ABC = 50°$ and $\angle DBC = 130°$. What can you say about ABD?

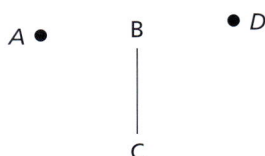

7 Measure this bearing.

N

8 Alan and Brian are standing in a field. The bearing of Alan from Brian is 132°. What is the bearing of Brian from Alan?

9 Make a copy of the diagram below, and on it mark out a bearing of 341° from point *X*.

N

X

10 Make a copy of the diagram below. Jamie starts from point *X* and walks on a bearing of 056°. Eriko starts from point *Y* and walks on a bearing of 310°. Mark the point where their paths cross.

X ●   ● *Y*

Exercise 2B

1 There are three angles below. Copy them, and label each of them with the word *acute*, *obtuse* or *reflex*.

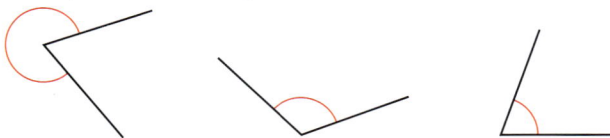

2 Use your protractor to draw an angle of 284°.

3 Calculate angle *a*.

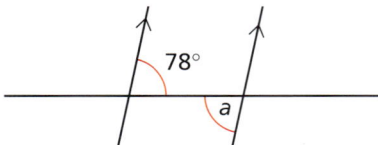

78°

a

4 Copy the diagram below and write in all of the other angles.

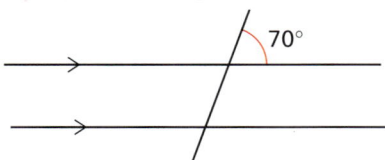

70°

> **Remember:**
>
> **Supplementary** angles add up to 180°. If two angles make up a straight line, and you know one angle, you can work out the other.

5 A bearing of angle *x* is between East and South. What can you say about *x*?

6 Give the bearing equivalent to NNW.
7 Measure this bearing.

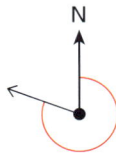

8 The bearing of Henry from Lionel is 322°. What is the bearing of Lionel from Henry?
9 Make a copy of the diagram below, and on it mark out a bearing of 129° from point X.

10 We have had several questions in which you found your position on a map. Is this always possible? If you can find the bearings of two distant objects, can you always find your position?

Exercise 2C Ma1

The instrument you use to find bearings is a *compass*. It has a magnetised needle which always points to North. You can make a compass from a disc of thin cardboard and a magnet. Draw a circle on thin card and cut it out. Mark the eight compass points round the disc. Glue the magnet to the disc, and balance it on a large drawing pin.

But you must be very exact to make sure that it balances at its centre!

You will need:
- thin cardboard
- compasses
- scissors
- pen
- glue
- thin bar magnet, or magnetised needle
- drawing pin

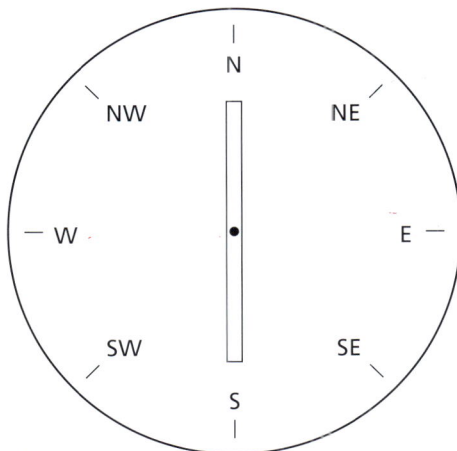

Exercise 2D Ma1

You don't always carry a compass! But provided you have a watch, and the sun is shining, you can find out roughly where North is, by this method.

- Take half-way between the hour hand of your watch and 12.
- Point this towards the sun.
- Then 12 itself will be pointing South, so the opposite is North.

This method works for Greenwich Mean Time (GMT), not for British Summer Time (BST). So, in the summer, subtract 1 hour from the time shown on your watch.

Can you explain why this works? Would it work in Australia? How could you make it work if you were living on the equator?

3 Percentages

'I never could make out what those damned dots meant.'

So said Lord Randolph Churchill, father of Sir Winston Churchill, evidently proud of his inability to understand decimal points.

There have always been people who find decimals and fractions difficult to handle. Percentages are another way of writing decimals and fractions, which some people find easier to work with.

Percentages, decimals and fractions

Decimals and fractions can be written in terms of **percentages**. One per cent is one part in 100. So there is an easy connection between percentages, fractions and decimals.

> **Percentage:** from the Latin *per centum* meaning 'by the hundred'.

1 per cent is the same as $\frac{1}{100}$ or 0.01.
2 per cent is the same as $\frac{2}{100}$ or 0.02.
3 per cent is the same as $\frac{3}{100}$ or 0.03.
10 per cent is the same as $\frac{10}{100}$, or 0.10.

A special sign, %, is used for *per cent*. We write 1 per cent as 1%.

20% is the same as $\frac{20}{100}$, or 0.20.
43% is the same as $\frac{43}{100}$, or 0.43.

And so on.

To convert a percentage to a decimal, divide by 100. Keep the decimal point fixed, and move the numbers two places to the right.

4	3	·		
		·	4	3

÷ 100

Be sure to put in a 0 if the percentage is less than 10.

7	·		
	·	0	7

÷ 100

Exercise 3.1

Change these percentages to decimals.

1 30%	**4** 73%	**7** 50%	**9** 3%
2 46%	**5** 99%	**8** 5%	**10** 12.5%
3 28%	**6** 40%		

It is just as easy to change from a decimal to a percentage: simply multiply by 100.

> 0.59 is the same as 59%.   0.09 is the same as 9%.
> 0.9 is the same as 90%.

You can do the change by moving the digits two places to the left.

	0	·	5	9
5	9	·		

× 100

Exercise 3.2

Change these decimals to percentages.

1 0.61	**4** 0.36	**7** 0.05	**9** 0.175
2 0.92	**5** 0.6	**8** 0.03	**10** 0.015
3 0.63	**6** 0.9		

Percentages are very useful when dealing with money, especially when working out discounts. You don't want to spend more than you have to! Often the prices of things are reduced so that they can be sold quickly.

Percentage discount

The rate of discount is usually given as a **percentage**. There are 100p in £1, so a discount of 20p in the pound is a discount of 20%.

Sale! 20% off – everything must go!

> A discount of 9p in the pound is 9%.
> A discount of 7p in the pound is 7%.
> A discount of 30p in the pound is 30%.
> A discount of 3p in the pound is 3%.

Examples What is the reduction on a £6 cassette, at a discount of 20%?
What is the sale price of the cassette after discount?

The discount is 20%, which is 20p in the £. The reduction is 6 × 20p, which is £1.20.
   So after the discount the price is £6 − £1.20, which is £4.80.

After a 15% discount, what is the sale price of a skirt which was originally £26?

The reduction is 26 × 15p, which is £3.90. So the price after discount is £26 − £3.90, which is £22.10.

Exercise 3.3

For these questions, calculate the reduction and the sale price after the discount.

1 A £2 cake, after a discount of 20%
2 A £5 book, after a discount of 10%
3 A £20 blouse, after a 10% discount
4 A £30 jacket, after a 20% discount
5 A £5 meal, after a 7% discount

6 A £30 fan heater, after a 20% discount
7 A £48 swimsuit, after a 15% discount
8 A £700 holiday, after a 10% discount
9 A £12 000 car, after a 12% discount
10 An 80p ice cream, after a 10% discount

If you are told the price before and after the reduction, you can find the discount. Each £ (100p) is reduced by a certain number of pence. This is the percentage discount.

Examples   A £1 cake is reduced to 90p. What is the percentage discount?

The price is reduced by 10p. This means that the discount is 10p in the £, which is $\frac{10}{100}$.
   The percentage discount is 10%.

A £15 tie is reduced to £12. What is the percentage discount?

The price is reduced by £3. So the reduction is £3 for £15 and the reduction per £ is £3 ÷ 15, which is £0.2, or 20p.

The percentage discount is 20%.

Exercise 3.4

For these questions, calculate the percentage discount.

1 A £1 drink reduced to 90p
2 A £1 card reduced to 80p
3 A £1 magazine reduced to 75p
4 A £1 calendar reduced to 60p
5 A £1 diary reduced to 30p
6 A £10 tie reduced to £9
7 A £20 artbook reduced to £15
8 A £50 computer game reduced to £40
9 A £200 suit reduced to £160
10 A £10 000 car reduced to £8000

Exercise 3.5   Ma 1

Go to your local shops, to see whether there is a sale in any of them. What is the discount on the goods in the sale? Make a note of the different ways the discount is shown.

Percentages on a calculator

When using a calculator for problems involving percentages, convert the percentages into decimal form.

Example   A compact disc costing £8.80 is sold at a discount of 15%. What is the reduction in price? What is the sale price?

Convert 15% to 0.15. Multiply £8.80 by 0.15, giving
£8.80 × 0.15 = £1.32.

So the reduction is £1.32. The sale price is £8.80 − £1.32 = £7.48.

Exercise 3.6

For these questions, put the percentage into decimal form, then use a calculator.

> **Remember:**
>
> *To convert a percentage to a decimal, divide by 100. To convert a decimal to a percentage, multiply by 100.*

1 A £46 skirt is sold at a discount of 31%. What is the reduction?

2 A £56 pair of shoes is discounted at 26%. What is the reduction?

3 A £11 500 car is discounted at 5%. What is the reduction? What is the sale price?

4 A £910 computer is discounted at 8%. What is the reduction? What is the sale price?

5 A package holiday costs £750. The tour operator offers a 15% discount if I pay immediately. What is the reduction? What is the reduced price?

6 A club offers a discount of 12% on its membership fee of £180, provided that it is paid within one month. What is the reduction? What is the reduced price?

Percentages of other quantities

You can find percentages of any quantity – it needn't be money.

Examples

A type of mortar contains 65% sand. How much sand is there in 28 kg of mortar?

Convert the percentage to a decimal, 0.65.

$28 \times 0.65 \text{ kg} = 18.2 \text{ kg}$

There is 18.2 kg of sand.

A carton of milk contains 400 ml, of which 5% is fat. How much fat is there in the carton?

5% of 400 ml is 400×0.05 ml, which is 20 ml.
There is 20 ml of fat.

After washing, a curtain lost $12\frac{1}{2}$% of its length. If it was 2 m long before washing, how long is it after washing?

$12\frac{1}{2}$% is the same as 12.5%. Convert to a decimal, 0.125. Multiply this by 2 m.

$$2 \times 0.125\,\text{m} = 0.25\,\text{m}$$

So 0.25 m was lost. The curtain is 1.75 m long after washing.

Remember:

Units of measurement include **kilograms (kg), millilitres (ml)** and **metres (m).**

Exercise 3.7

Find the values of the following, using a calculator. Write the percentages in decimal form, and remember the units of measurement.

1 23% of 200 cm
2 47% of 1200 kg
3 35% of 43 km
4 60% of 45 cm
5 6% of 45 cm
6 3% of 150 kg
7 $12\frac{1}{2}$% of 80 grams
8 A forest consists of 10 500 trees. In a fire, 27% are destroyed. How many trees are destroyed?
9 A field contains 5000 tulips, of which 29% are yellow. How many yellow tulips are there?
10 A laser printer costs £480, before VAT (Value Added Tax). How much is the VAT, at a rate of $17\frac{1}{2}$%?

Percentages as fractions, fractions as percentages

The metric and decimal systems are used widely, and many people use a calculator to do most of their arithmetic, so decimals are often more useful than fractions. However, a percentage is just a fraction with a denominator of 100, and there is an easy connection between percentages and fractions.

$$1\% = \tfrac{1}{100} \qquad 2\% = \tfrac{2}{100} \qquad 5\% = \tfrac{5}{100}$$
$$10\% = \tfrac{10}{100} \qquad 27\% = \tfrac{27}{100} \qquad 50\% = \tfrac{50}{100}$$

Some of these fractions can be simplified.

Learn these by heart

$$50\% = \tfrac{50}{100} = \tfrac{1}{2}$$
$$25\% = \tfrac{25}{100} = \tfrac{1}{4}$$
$$75\% = \tfrac{75}{100} = \tfrac{3}{4}$$
$$10\% = \tfrac{10}{100} = \tfrac{1}{10}$$
$$5\% = \tfrac{5}{100} = \tfrac{1}{20}$$
$$33\tfrac{1}{3}\% = \tfrac{33\frac{1}{3}}{100} = \tfrac{1}{3}$$
$$66\tfrac{2}{3}\% = \tfrac{66\frac{2}{3}}{100} = \tfrac{2}{3}$$

Examples   Change 40% to a fraction. Simplify your answer as far as possible.

Divide 40 by 100. $40\% = \frac{40}{100} = \frac{4}{10} = \frac{2}{5}$.

Change $\frac{4}{5}$ to a percentage.

Multiply $\frac{4}{5}$ by 100. $\frac{4}{5} \times 100 = \frac{400}{5} = 80$. So the percentage is 80%.

Exercise 3.8

Change these percentages to fractions. Simplify your answers as far as possible.

1 30%
2 20%
3 3%
4 6%
5 15%

> **Remember:**
> *To convert a percentage to a fraction, divide by 100. To convert a fraction to a percentage, multiply by 100.*

Changes these fractions to percentages.

6 $\frac{2}{5}$
7 $\frac{3}{5}$
8 $\frac{9}{10}$
9 $\frac{13}{20}$
10 $\frac{1}{8}$

You could keep a record of your own progress in tests. Change all your scores to percentages, and you can see how much you have improved.

Example   For his Geography tests, Leroy got 56 marks last week and 45 marks this week. But the first test was out of 80 and the second out of 60. Which test did he score higher in?

Change both marks into percentages to compare them.
For the first test, $\frac{56}{80} = \frac{7}{10} = 70\%$.
For the second test, $\frac{45}{60} = \frac{3}{4} = 75\%$.
So his percentage score was higher in the second test.
Using a calculator, the results are as follows.

$$56 \div 80 = 0.7 = 70\% \qquad 45 \div 60 = 0.75 = 75\%$$

Exercise 3.9

In these questions, work out the pairs of test scores as percentages, and say which is higher.

	first test	second test
1	$\frac{27}{30}$	$\frac{60}{80}$
2	$\frac{41}{50}$	$\frac{36}{45}$
3	$\frac{28}{40}$	$\frac{36}{60}$
4	$\frac{17}{20}$	$\frac{18}{30}$
5	$\frac{27}{90}$	$\frac{27}{80}$

One quantity as a percentage of another

In exercise 3.9 we converted test scores to a percentage of the total marks available. This sort of calculation can be used more generally.

Examples

A shopkeeper buys a lamp for £40 and sells it for £50. Calculate her percentage profit.

The actual profit is £10. Divide this by 40 (the buying price was £40), then convert to a percentage.

$$10 \div 40 = \tfrac{10}{40} = 0.25 = 25\%$$

So the percentage profit is 25%.

The next week the shopkeeper buys a similar lamp for £50, and sells it for £60. Calculate her percentage profit.

Again the actual profit is £10. Divide by 50, then convert to a percentage.

$$10 \div 50 = \tfrac{10}{50} = 0.2 = 20\%$$

Using percentages, we see immediately that the shopkeeper's rate of profit is reduced.

● *In general, to find one quantity as a percentage of another, divide the first by the second, then multiply by 100 to convert to a percentage.*

Exercise 3.10

Find the first quantity as a percentage of the second.

1 7 as a percentage of 14
2 30 as a percentage of 50
3 £3 as a percentage of £15
4 £6.84 as a percentage of £19
5 4 pints as a percentage of 5 pints

6 34p as a percentage of £1.36
7 15 cm as a percentage of 1 m
8 800 m as a percentage of 1 km
9 1600 g as a percentage of 5 kg
10 12.5 kg as a percentage of 50 kg

Exercise 3.11

1 A secondhand book seller bought a book for 30p and sold it for 36p. What was her percentage profit? What should she sell the book for, to make a 40% profit?
2 A chair is bought for £250 and sold for £290. What is the percentage profit? What price should it be sold at to make a profit of 20%?
3 An investor buys shares for £1200, and later sells them for £1500. What is the percentage profit?

4 An investor buys shares for £16 000, but has to sell them for £14 000. What is the percentage loss?
5 Which of the following transactions gives the greater percentage profit?
   a Buying for £13 and selling at £14.82
   b Buying for £12 and selling at £13.56.
6 Wally buys a consignment of mugs for £2500, and aims to make a 15% profit. He sells half of the mugs at 15% profit, but then has to sell the rest of the mugs at only 10% profit. What was his profit for the transaction as a whole?
7 Rajiv bought 350 diaries at 57p each. He sold 150 of the diaries at 80p each, but had to sell the rest at only 50p each. What was his percentage profit or loss for the transaction as a whole?

Multiplying factor for decrease

In this chapter we have solved many problems involving discount. To find the price after a discount, find the reduction and then subtract it from the original price.

There are two arithmetical operations here, multiplication and subtraction. We can do both of these operations in one go, using a **multiplying factor**.

Suppose there is a discount of 20% on all items. Then 20% of the price has been removed, leaving 80% of the price. The sale price is 80% of the original price. To find the sale price, multiply the original price by 0.8.

Examples A suit cost £90 before a 20% discount. What is its price after the discount?

If 20% is deducted, 80% remains. The price after the discount is 80% of £90, so multiply the price by 0.8.

$$0.8 \times 90 = 72$$

The price after discount is £72.

The area of forest in a country decreases by 5%. If the area was 120 000 square miles before the decrease, what is it after the decrease?

If 5% is lost, then 95% remains. The percentage 95% is equivalent to the decimal 0.95. Multiply 120 000 by 0.95.

$$0.95 \times 120\,000 = 114\,000$$

So 114 000 square miles of forest are left.

In general, suppose you are given a percentage decrease. Subtract this from 100 to find the percentage which remains. Convert to a decimal by dividing by 100. This gives the multiplying factor.

> **Multiplying factor (decrease)**
> Percentage remaining = 100 − percentage decrease
> Multiplying factor = percentage remaining ÷ 100

Exercise 3.12

In questions 1 to 5 find the multiplying factor corresponding to the percentage decrease.

1 10%
2 30%
3 3%
4 8%
5 50%
6 What is the multiplying factor corresponding to a 10% discount? If a pair of shoes costs £40 before the discount, what does it cost after?
7 The population of a town decreases by 5%. What is the multiplying factor corresponding to this decrease? If the population was 400000 before the decrease, what is it after?
8 A boxer goes on a diet and reduces his weight by 15%. What is the multiplying factor corresponding to this decrease? If he weighed 10 stone before the diet, what does he weigh after?
9 In a drought, the diameter of a baobab tree decreased by 5%. If it was 160 cm before the drought, what was it after?
10 Wet firewood was put in a barn to dry. Over a week, its mass decreased by 2%. If there was 200 kg before the decrease, how much was there after?

Multiplying factor for increase

We saw how the multiplying factor method makes calculations of decrease easier. You use one operation instead of two. This also works for calculations of *increase*.

Examples   Your money increases by 10%. What is the multiplying factor?

For every £100 you had originally, you now have £110. Your money has been multiplied by 110/100, that is, by 1.1.

The weight of a cow increases by 5%. What is the multiplying factor?

For every 100 kg the cow weighed originally, it now weighs 105 kg. The cow's weight has been multiplied by 105/100, that is, by 1.05.

In general, add the percentage increase on to 100, then convert to a decimal by dividing by 100. This gives the multiplying factor.

> **Multiplying factor (increase)**
> Multiplying factor = (percentage increase + 100) ÷ 100

Exercise 3.13

In questions 1 to 5 find the multiplying factor corresponding to the percentage increase.

1 30%
2 50%
3 4%

4 12%
5 100%

6 What is the multiplying factor corresponding to an increase of 20%? A holiday was priced at £340, but then the price went up by 20%. What is the increased price?

7 What is the multiplying factor corresponding to an increase of 15%? Over the Christmas period, George's weight increased by 15%. If he weighed 70 kg before Christmas, how much did he weigh after?

8 Thanks to a new fertiliser, the yield of a cereal farm increases by 10%. If it produced 3500 tonnes before, how much does it produce after?

9 The population of a country is increasing at 4% each year. If it is 12 000 000 now, what will it be in a year's time?

10 A clever stockbroker increases the value of his investments by 150%. If they used to be worth £120 000, what are they worth after the increase?

SUMMARY

■ A percentage is a way of writing fractions or decimals. The expression 1% is the same as $\frac{1}{100}$, or 0.01.

■ You can convert between decimals and percentages, by dividing or multiplying by 100. You should know the percentages equivalent to certain fractions, such as 50% and $\frac{1}{2}$.

■ During sales, prices are discounted. Discounts can be expressed as percentages. A 20% discount is equivalent to a reduction of 20p in the £.

■ Percentages can also express increases. A price rise of 20% means that a price of £100 increases to £120.

■ You can express one quantity as a percentage of another.

■ Using a multiplying factor is an efficient way to deal with percentage decrease or increase.

Exercise 3A

1 What is the reduction in the price of a £40 dress, after a 20% discount?
2 What is the price of a £600 computer after a 15% discount?
3 **a** Convert 0.02 to a percentage.
  b Convert 74% to a decimal.
4 **a** Convert $\frac{7}{8}$ to a percentage.
  b Convert 35% to a fraction in its simplest form.
5 A bottle of gin contains 40% alcohol. If there is 700 ml of liquid in the bottle, how much alcohol is there?
6 A school of 600 pupils contains 318 girls. What percentage of the school pupils are girls?
7 Which would give the larger percentage profit: a chair bought for £200 and sold for £250, or a table bought for £500 and sold for £620?
8 A salary of £19 000 is increased by 5%. What is the new salary?
9 What is the multiplying factor corresponding to a decrease of 25%? An air journey time has been cut by 25%. If the journey used to take 8 hours, find the time it now takes.
10 What is the multiplying factor corresponding to an increase of 12%? Find the cost of a £150 coat after its price has been increased by 12%.

Exercise 3B

1 What is the price of a £15 000 car after a 20% discount?
2 A £300 television is reduced to £270. What is the percentage discount?
3 **a** Convert 45% to a decimal.
  b Convert 0.12 to a percentage.
4 A salary is increased to 'half as much again'. What is this as a percentage increase?
5 Convert $33\frac{1}{3}$% to a fraction.
6 In a factory, 35% of the employees are women. How many women are there if there are 700 employees?
7 A breakfast cereal contains 5 grams of fibre in each 50 grams. What percentage of the cereal is fibre?
8 Mr Hussain's season ticket price is cut from £120 to £108. Ms Johnson's season ticket price is cut from £160 to £140. Who has had the greater percentage cut?
9 In training, a jockey reduces his weight by 5%. Before training he weighed 120 pounds. Use the multiplying factor method to find his new weight.
10 One year, the total rainfall recorded was 88 cm. The next year it increased by 15%. Use the multiplying factor method to find the increased rainfall.

Exercise 3C · Ma1

Collect information from various sources which uses percentages in some way. Listen to the news on radio or television and write down any facts involving percentages. You might list the information under headings like inflation, employment, wages, housing, tax, stock market, education, health service, mortgages, cost-of-living, etc.

After doing this for a week, make a class poster of the general uses of percentages. Discuss your results.

Exercise 3D · Ma1

At present the rate of VAT is $17\frac{1}{2}$%. What an awkward number!

Suppose a computer game costs £40 before VAT. How much is the VAT? (Use a calculator.)

There is a way to find the VAT without a calculator.

- Take a tenth of £40.   £4
- Take half of this amount.   £2
- Take half of this amount.   £1

Add these amounts, obtaining £7. It should be the same as your previous answer!

Test this method on some other things you want to buy. Does it always work? Can you explain why it works?

4 Quadrilaterals and polygons

Brian spilled ink on a tablecloth. Fortunately there was some extra material that matched the tablecloth. Brian cut out a rectangle from the tablecloth, and an equal rectangle from the extra material. He made sure the pattern matched.

Brian sewed the rectangle into the hole. His sister Kate was scornful.

How many ways could Brian sew the rectangle into the hole?

What if Brian had cut out a square hole and a square piece of cloth: how many ways could the square piece be sewed into the square hole?

> **Remember:**
> *A rectangle has all angles equal to 90°. A square has all angles equal to 90° and all sides equal.*

A square is a special sort of rectangle. It is the top-of-the-range rectangle.

Types of quadrilateral

Quadrilaterals you should know

- Rectangle: all angles are equal to 90°

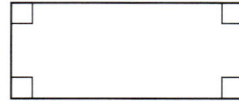

- Square: all sides are equal; all angles are equal to 90°

- Rhombus: all sides are equal

- Parallelogram: opposite sides are parallel

- Kite: two pairs of adjacent sides are equal

- Trapezium: one pair of opposite sides are parallel

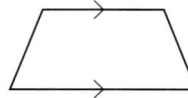

Squares and rectangles are all around us, in the shapes of doors, sheets of paper, table tops, and so on. The other types of quadrilateral are less common, but you can find them in some everyday objects.

Rhombus   The diamonds in a pack of cards are rhombuses.
Parallelogram   Take a stack of paper, and look at it from one end. You see a rectangle. Now push the stack slightly at the top. You see a parallelogram.
Kite   The kite that you fly on the end of a string is, of course, a kite.

Look, there's a kite in the sky!

But that isn't a kite! This is one.

Trapezium   The side window of a car is a trapezium, with parallel horizontal sides. The side of a shed, or the side wall of a swimming bath, might be a trapezium, with parallel vertical sides.

Exercise 4.1

1 Name the quadrilaterals shown below.

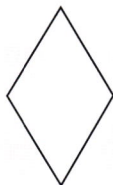

2 The diagram below shows a hexagon, *ABCDEF*, with all sides equal and all angles equal. The centre is at *O*. What are the names for the following quadrilaterals?

a *ABCF*
b *ABDE*
c *ABOF*
d *ABCE*

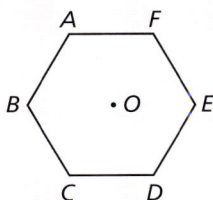

3 A shape is both a rectangle and a rhombus. What type of quadrilateral is it?
4 A shape is both a parallelogram and a kite. What type of quadrilateral is it?
5 What is the name of a quadrilateral that is both a kite and a rectangle?
6 Suppose four equal strips of cardboard were joined end to end as shown below. What sort of quadrilateral would be made?

7 Suppose two equal equilateral triangles were joined edge to edge as shown below. What sort of quadrilateral would be made?

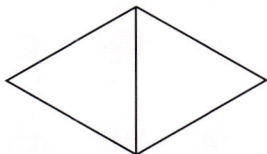

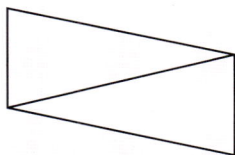

8 Suppose two equal isosceles triangles were joined edge to edge as shown below. What sort of quadrilateral would be made?

9 Suppose two *different* isosceles triangles have the same base side. What sort of quadrilateral would be made if they were joined by their base sides, as shown.

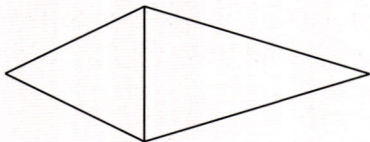

10 What sort of quadrilaterals can be made by joining two identical scalene triangles?

Other properties of quadrilaterals

These special sorts of quadrilateral often have many other properties. We defined a rectangle as a quadrilateral with all angles equal to 90°. The opposite sides of a rectangle are parallel, so a rectangle is a special sort of parallelogram.

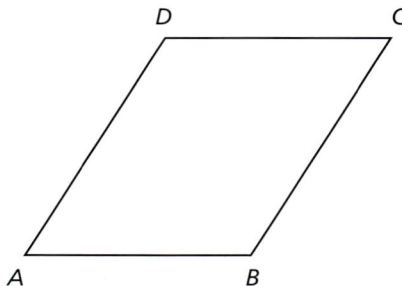

Look at the parallelogram $ABCD$. Measures the angles. You should find that the opposite angles are equal: $\angle BAD$ and $\angle BCD$ are both equal to 58° and $\angle ABC$ and $\angle ADC$ are both equal to 122°. Measure the sides of $ABCD$. You should find that opposite sides are equal: AB and CD are both equal to 3 cm, and BC and AD are both equal to 3.5 cm.

Copy the parallelogram $ABCD$. Join the diagonals AC and BD of the parallelogram. They meet at X. You should find that X cuts both AC and BD exactly in half.

In a parallelogram:
● *opposite angles are equal,*
● *opposite sides are equal,*
● *diagonals cut each other in half.*

Exercise 4.2
●●●●●●●●●

The diagram below shows a rhombus, *ABCD*, with all sides equal. Make a copy of the diagram and join the diagonals *AC* and *BD*. Measure the angles and the lengths within the figure. You should find the following results.

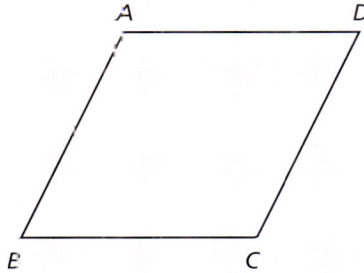

In a rhombus:
● *opposite angles are equal,*
● *opposite sides are parallel,*
● *the diagonals AC and BC cut each other in half, at 90°,*
● *the diagonal AC cuts ∠BAD in half.*

Exercise 4.3
●●●●●●●●●

The diagram below shows a rectangle, which has all angles equal to 90°. Make a copy of the diagram and join the diagonals *AC* and *BD*.

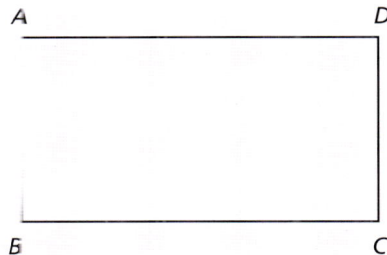

a Measure the sides and the diagonals of the rectangle. What do you notice?
b The diagonals cross at *X*, say. What can you say about *AX*, *BX*, *CX* and *DX*?
c Write down all the facts you have discovered about the rectangle.

Symmetry properties

At the beginning of this chapter Brian and Kate had the problem of putting a rectangle into a rectangular hole. There are four ways of doing this:

- the right way,
- upside down,
- back to front,
- upside down *and* back to front.

> **Remember:**
>
> *A **line of symmetry** is a fold line which cuts a figure in half with both halves matching exactly.*
>
> *A figure has **rotational symmetry** if it can be rotated by less than 360° and cover its own shape exactly.*

This is because a rectangle has two **lines of symmetry**, and **rotational symmetry**.

The lines of symmetry are the lines through the centres of opposite sides. The two halves of the rectangle are the same on both sides of each line of symmetry.

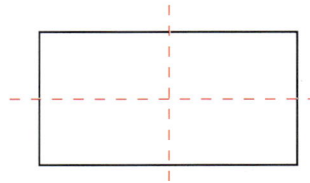

If you rotate the rectangle 180° about the central point, it covers its shape. If you rotated through another 180° it would be back where it started. Hence the rectangle has rotational symmetry of order 2.

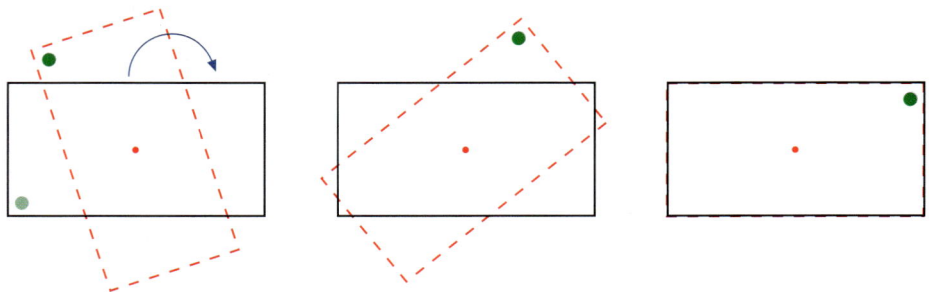

Exercise 4.4

What is the connection between the symmetry of the rectangle and the ways in which it can be put into a rectangular hole? Rotational symmetry corresponds to putting the rectangle in upside down. What corresponds to the lines of symmetry?

Exercise 4.5   Ma1

Cut a square out of a piece of paper.

1 In how many ways can you put the square into the square hole?
2 How many lines of symmetry does the square have?
3 What is its rotational symmetry?

You will need:
• paper
• ruler
• pencil
• scissors

Exercise 4.6   Ma1

Repeat exercise 4.5 for a rhombus and a kite. Write up what you have found about the symmetry of these figures.

Exercise 4.7

In questions 1 to 8, copy the statements and replace the blanks with one of the words *square*, *rectangle*, *rhombus*, *parallelogram*, *kite* or *trapezium*. In some questions there is more than one correct answer.

1 A _____ has all angles equal.
2 A figure with two pairs of equal sides and one line of symmetry is a _____ .
3 A figure with two pairs of equal sides and no lines of symmetry is a _____ .
4 A figure with rotational symmetry of order 4 is a _____ .
5 A figure with equal sides and two lines of symmetry is a _____ .
6 A _____ has four lines of symmetry.
7 Every _____ is also a rectangle.
8 Every parallelogram is also a _____ .
9 Describe three properties that a rhombus has but a parallelogram may not have.
10 Describe three properties that a square has but a rectangle may not have.

Angles of a quadrilateral

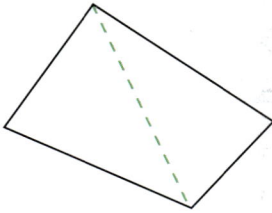

A rectangle has four angles, each equal to 90°. Hence the sum of its angles is $4 \times 90°$, that is, 360°.

In fact, the sum of the angles of *any* quadrilateral is 360°. You can divide *any* quadrilateral into two triangles, as shown on the left. The angles in each triangle add up to 180°, so the angles in a quadrilateral add up to $2 \times 180°$.

● *The sum of the angles of a quadrilateral is 360°.*

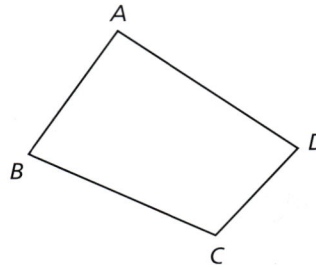

In quadrilateral $ABCD$,
$\angle BAD + \angle CBA + \angle DCB + \angle ADC = 360°$.

● *If you know three of the angles of a quadrilateral, then you can find the fourth.*

Examples   In the quadrilateral below, $\angle BAD = 79°$, $\angle ABC = 108°$ and $\angle BCD = 111°$. Find $\angle ADC$.

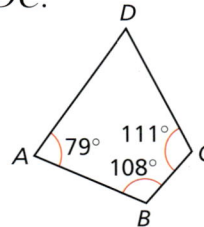

The sum of all four angles is 360°. So, to find $\angle ADC$, subtract the other three angles from 360°.

$$360° - 79° - 108° - 111° = 62°.$$

So $\angle ADC$ is equal to 62°.

In the quadrilateral below, $\angle BAD = 87°$ and $\angle ABC = 131°$. $\angle BCD$ and $\angle ADC$ are equal to each other. Find $\angle BCD$.

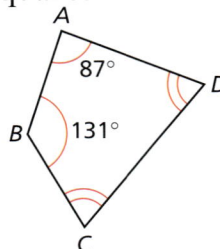

To find the sum of ∠*BCD* and ∠*ADC*, subtract the other two angles from 360°.

∠*BCD* + ∠*ADC* = 360° − 87° − 131° = 142°.

As ∠*BCD* and ∠*ADC* are equal they are equal to half of 142°.

∠*BCD* = ½ × 142° = 71°.

So ∠*BCD* is equal to 71°.

Exercise 4.8

In questions 1 to 4 calculate the unknown angles in the quadrilaterals.

1

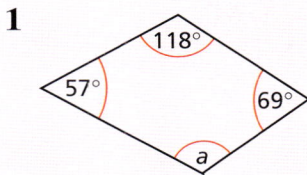

3

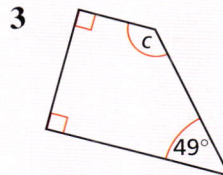

2

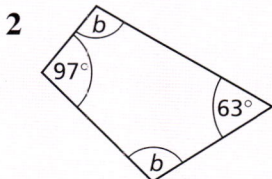

4

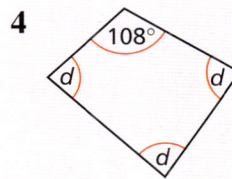

5 The diagram below shows a rhombus *ABCD*. ∠*DAC* = 30°. Find ∠*ADC*.

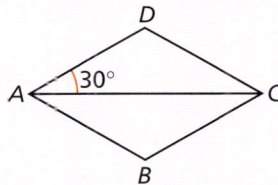

6 The diagram below shows a kite *ABCD*, with *AB* = *AD* and *CB* = *CD*. ∠*BAD* = 70° and ∠*ABC* = 130°. Find ∠*BDC*.

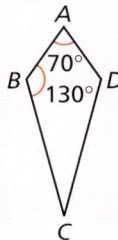

7 In the quadrilateral shown below, *a* + *c* = 90°. Find the value of *b* + *d*.

8 *ABCD* is a square, whose diagonals meet at *E*. Find ∠*EAB*.

9 *ABCD* below is a rhombus. The diagonals meet at *E*. ∠*BAE* = 40°. Find ∠*EDC*.

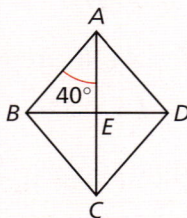

10 *ABCD* below is a kite, with *AB* = *AD* and *CB* = *CD*. The diagonals meet at *E*. ∠*BAE* = 60° and ∠*ECD* = 20°. Find ∠*CDA*.

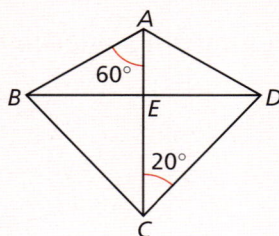

Polygons

A triangle has three straight sides. A quadrilateral has four straight sides. These are **polygons**. Do you remember the names for the next few polygons?

Exercise 4.9

1 What is the name for a five-sided polygon?
2 What is the name for a six-sided polygon?
3 What is the name for an eight-sided polygon? (Hint: what sea creature has eight tentacles?)
4 The athletic event with seven components is the *heptathlon*. What is the name for a seven-sided polygon?
5 The athletic event with ten components is the *decathlon*. What is the name for a ten-sided polygon?

> ● *A polygon is **regular** if all its sides are equal and all its angles are equal.*

A regular triangle is **equilateral**. A regular quadrilateral is a **square**.

You have already met regular triangles and quadrilaterals.

The diagram at the top of page 59 shows a regular pentagon, a regular hexagon and a regular heptagon (seven-sided).

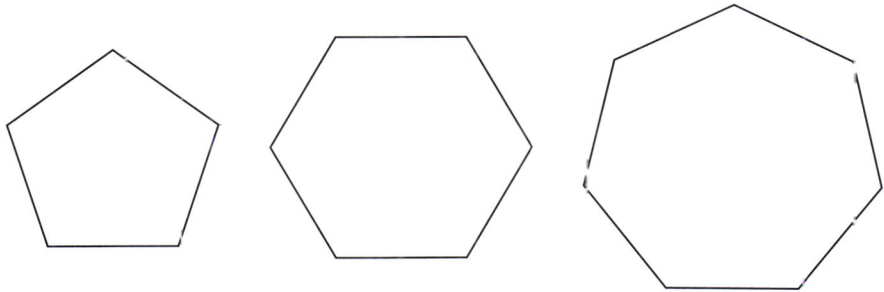

We already know that the sum of the angles of a triangle is 180°, and that the sum of the angles of a quadrilateral is 360°. What about the next few polygons? In this exercise you will find out.

Exercise 4.10

1 The diagram below shows a pentagon, *ABCDE*, with *A* joined to *D* and to *C*.

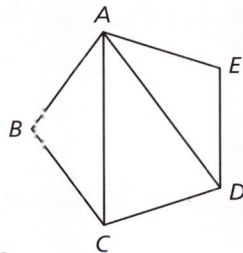

a How many triangles are there?
b What is the sum of the angles of *ABCDE*?
2 The diagram below shows a hexagon, *ABCDEF*, with *A* joined to *D*, *E* and *C*.

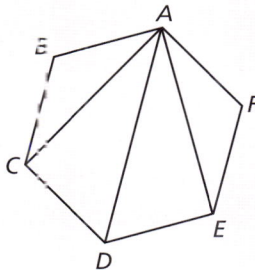

a How many triangles are there?
b What is the sum of the angles of *ABCDEF*?
3 Apply the method of questions 1 and 2 to a heptagon.
  a How many triangles do you get?
  b What is the sum of the angles of a heptagon?
4 What is the sum of the angles of an octagon (eight sides)?
5 A dodecagon has 12 sides. What is the sum of its angles?

Have you found a pattern?
  If you increase the number of sides by 1, you add one triangle, and so you increase the sum of the angles by 180°.

Exercise 4.11

1 Copy and complete this table.

number of sides	3	4	5	6	7	8	9
number of triangles	1	2	3				
sum of angles	180°	360°					

Notice that the number of triangles is 2 less than the number of sides. Can you discover a rule or method to find the sum of the angles?

2 A pentagon has 5 sides. Check that your method works for a pentagon.
3 An icosagon has 20 sides. Use your rule to find the sum of its angles.
4 What is the sum of the angles of a 100-sided polygon?
5 If a polygon has *n* sides, what is the sum of its angles?

● *The rule to find the sum of the angles of a polygon is: subtract 2 from the number of sides, then multiply by 180°.*

Was your rule similar to this?

Tessellations

Patterns are made up from basic shapes used over and over again. If the shapes fit together without gaps, the pattern is a **tessellation**.

What basic shapes could we use? Obviously, we could use squares. The tiling of a kitchen or bathroom wall is usually from square tiles.

You can also tessellate with rectangles. This shows the easiest way to fit the rectangles together.

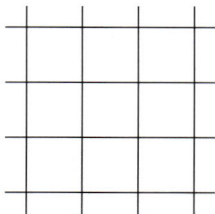

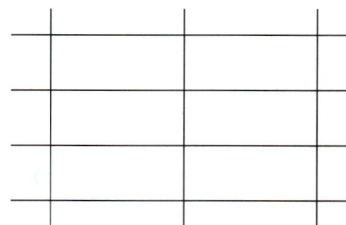

Floor tiles are often rectangular. In particular, *parquet flooring* consists of rectangular blocks of wood called *parquets*. These are not fitted together in the simple way shown above.

Exercise 4.12 Ma1

Look at the floors in the places you visit, and note when they have rectangular tiles, in particular when they have parquet flooring.

1 What patterns are formed?
2 Are they always laid in the same way?

Exercise 4.13 Ma1

When you are walking around, look at the patterns of bricks on houses. If there are some old houses in your neighbourhood, try to find them, as their brick patterns are often different. Some of the patterns you might find are *English bond*, *Flemish bond* and *herringbone*.

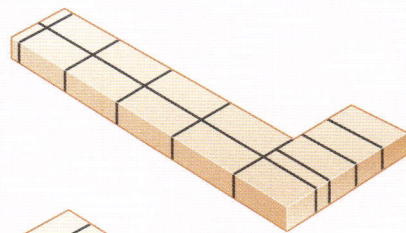

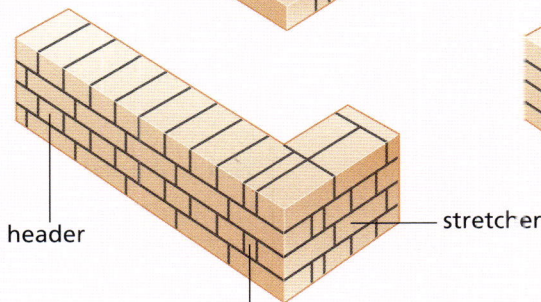

header   stretcher

English bond   closer

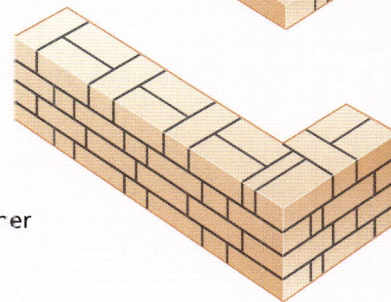

Flemish bond

herringbone

What about less regular quadrilaterals? Can you tile with rhombuses, for example?

Exercise 4.14

Make ten copies of the rhombus shown below, and cut them out. Show how you can arrange them in a tessellation.

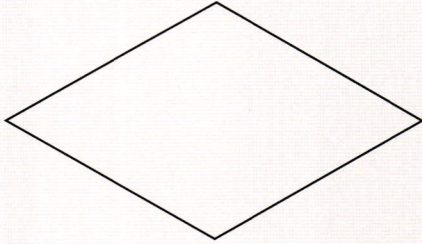

You can tessellate with *any* triangle.

Exercise 4.15

Make ten copies of the triangle below, and cut them out. Show how you can tessellate with them.

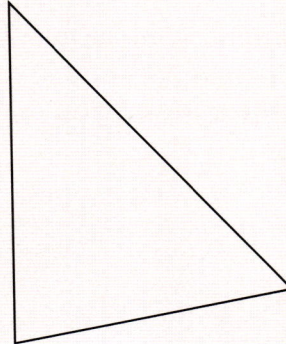

You can tessellate with triangles and with quadrilaterals. What about other polygons: pentagons, hexagons, and so on?

Exercise 4.16

1 Cut out several copies of the regular pentagon shown below. Can you tessellate with them?

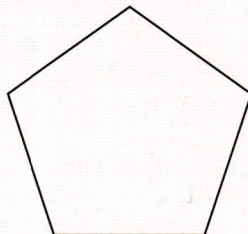

2 Cut out several copies of the regular hexagon shown below. Can you tessellate with them?

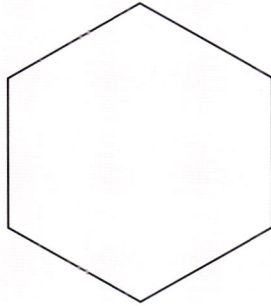

3 Cut out several copies of the regular heptagon (seven-sided polygon) shown below. Can you tessellate with them?

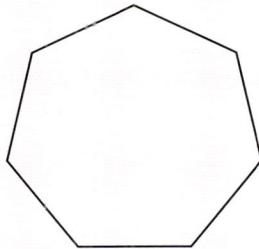

You should find that you can tessellate with hexagons, as well as with triangles and quadrilaterals, but not with the other regular polygons. Why is this so?

Exercise 4.17

In exercise 4.11 we found the sum of the angles of triangles, quadrilaterals, pentagons, and so on. If the polygon is regular, then the angles are equal. Copy and complete the following table.

number of sides	3	4	5	6	7	8	9
angle		90°		120°		135°	

For each possible internal angle, divide 360° by it. When is the result a whole number?

You should have found that the only regular polygons that will tessellate are triangles, quadrilaterals and hexagons.

SUMMARY

- ■ There are several special quadrilaterals.
- ■ The special quadrilaterals have special symmetry properties.
- ■ The sum of the angles of a quadrilateral is 360°. If you know three of the angles of a quadrilateral, then you can find the fourth.
- ■ A regular polygon has all its sides equal and all its angles equal. As the number of sides increases by 1, the sum of the angles increases by 180°. To find the sum of the angles of a polygon, subtract 2 from the number of sides, then multiply by 180°.
- ■ You can tessellate with triangles, quadrilaterals and hexagons, but not with any other regular polygon.

Exercise 4A

1 Name this polygon.

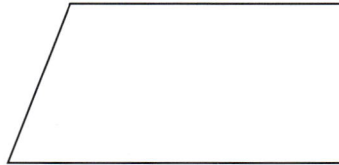

2 A rhombus *ABCD* is cut in half along *AC*. What sort of triangle is *ABC*?

3 A quadrilateral with a reflex angle is a *concave* quadrilateral. Can a concave quadrilateral be any of the special types of quadrilateral of this chapter?

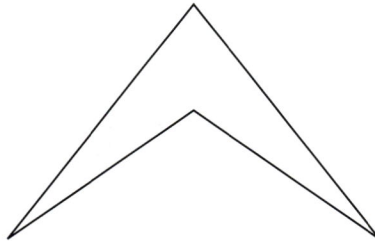

4 A trapezium in which the non-parallel sides are equal is an **isosceles trapezium**. What are the symmetries of an isosceles trapezium?

5 Three angles of a quadrilateral are 97°, 117° and 105°. What is the fourth angle?

6 Three angles of a quadrilateral are each 25°. What is the fourth angle? See if you can sketch this quadrilateral. What is its name?

7 What is the sum of the angles of a 15-sided polygon?

8 *ABCDE* is a regular five-sided polygon. How many lines of symmetry does it have? What is the order of its rotational symmetry?

9 Copy the right-angled triangle shown below. By drawing several more triangles, show that you can tessellate with the triangle.

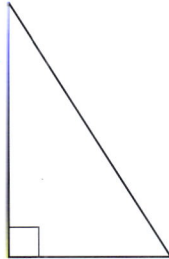

10 Cut out ten copies of the concave quadrilateral of question 3. Show how you can tessellate with them.

Exercise 4B

1 The triangle ABC below is cut by a line DE parallel to BC. What is the name of the quadrilateral $BCED$?

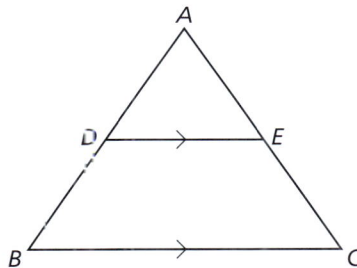

2 In a quadrilateral $ABCD$, $A = 66°$ and the other three angles are equal. What are the other angles?

3 $ABCD$ is a kite, with $AB = AC$ and $DB = DC$. What sort of triangles are ABC and DBC?

4 $ABCDE$ is a regular pentagon, and F is its centre. Find $\angle AFB$ and $\angle FAB$.

5 What is the sum of the angles of a 17-sided polygon?

6 $ABCDEF$ is a regular six-sided polygon. How many lines of symmetry does it have? What is its order of rotational symmetry?

7 $ABCD\ldots$ is a regular n-sided polygon. How many lines of symmetry does it have? What is its order of rotational symmetry?

8 If all the angles of a triangle are equal, then it is regular (equilateral). Is the same true for a quadrilateral?

9 If all the sides of a triangle are equal, then all the angles are equal. Is the same true for a quadrilateral?

10 *ABCDE* shows a (non-regular) pentagon. Show that you can tessellate with this shape.

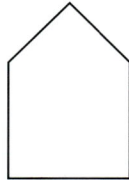

Exercise 4C

If you use only one type of polygon to tessellate, it has to be a triangle, quadrilateral or hexagon. But you can use other polygons in combination. Below are five tessellation patterns with more than one type of polygon. Try to describe them in terms of the polygons, the angles, and so on.

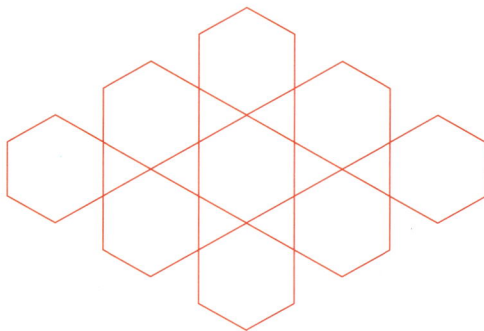

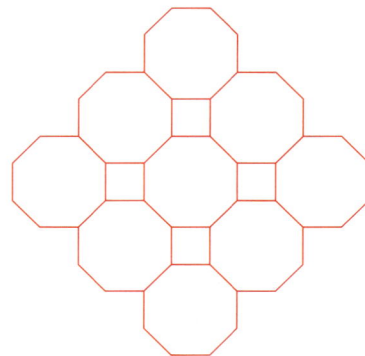

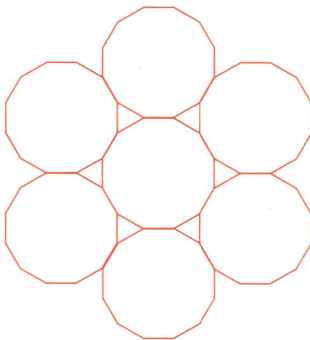

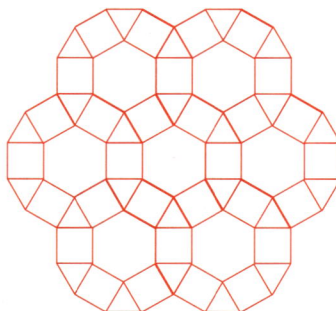

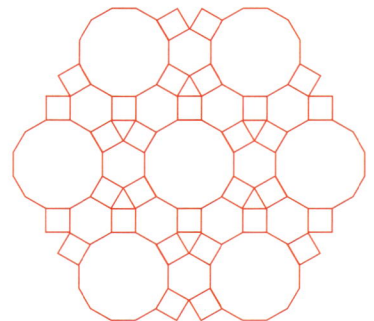

Exercise 4D

The angles inside a polygon are the **interior** angles. The angles outside it, shown below, are the **exterior** angles. How do we find the sum of these exterior angles?

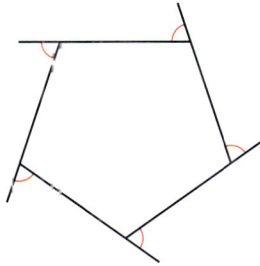

1   The diagram below shows a regular pentagon. What is the interior angle? What is the exterior angle *a*? What is the sum of the exterior angles, *a*, *b*, *c*, *d* and *e*?

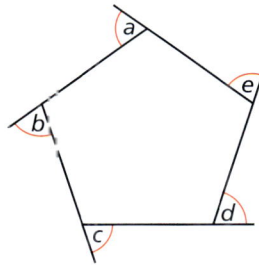

2   The diagram below shows a regular hexagon. What is the interior angle? What is the exterior angle *a*? What is the sum of these exterior angles, *a*, *b*, *c*, *d*, *e* and *f*?

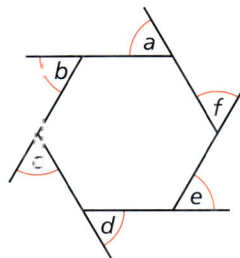

3   What is the sum of the exterior angles of a regular heptagon?
4   What is the sum of the exterior angles of a regular octagon?
5   *ABCDE* is a (non-regular) pentagon with interior angles 100°, 85°, 130°, 105° and 120°. Find the sum of the exterior angles.

What have you found about the sum of the exterior angles of a polygon? Can you justify it?

5 : Algebraic expressions

Anne, Beavis and Chloe are on a school skiing trip in Switzerland. They are looking at the prices in a gift shop, for presents to take home.

How can you write down a calculation, showing that you are doing the addition before the multiplication?

Order of operations

If you add several numbers together, it doesn't matter which numbers you add first.

$$5 + 6 + 8 = 5 + 14 = 19 \quad \text{(if we add 6 and 8 first, and then add the 5)}$$
$$5 + 6 + 8 = 11 + 8 = 19 \quad \text{(if we add 5 and 6 first, and then add the 8)}$$

Just imagine yourself at the checkout of a supermarket. It doesn't matter what order the prices are added together, the bill should come out to the same total!

Similarly, if you have several numbers multiplied together, it doesn't matter which numbers you multiply together first.

$$5 \times 6 \times 8 = 5 \times 48 = 240 \qquad \text{(if we multiply 6 and 8 first, then multiply by 5)}$$

$$5 \times 6 \times 8 = 30 \times 8 = 240 \qquad \text{(if we multiply 5 and 6 first, then multiply by 8)}$$

Just imagine yourself working out the capacity of a freezer which is 30 cm deep, 40 cm wide and 50 cm high. It doesn't matter which numbers you multiply together first!

Exercise 5.1

In questions 1 to 5, copy the equations and fill in the blanks. Make sure that you get the same answer whichever order you do the operations.

1 $3 + 4 + 5 = 3 + \underline{\quad} = \underline{\quad}$ $\qquad 3 + 4 + 5 = \underline{\quad} + 5 = \underline{\quad}$

2 $9 + 8 + 10 = 9 + \underline{\quad} = \underline{\quad}$ $\qquad 9 + 8 + 10 = \underline{\quad} + 10 = \underline{\quad}$

3 $12 + 13 + 17 = 12 + \underline{\quad} = \underline{\quad}$ $\qquad 12 + 13 + 17 = \underline{\quad} + 17 = \underline{\quad}$

4 $2 \times 3 \times 4 = 2 \times \underline{\quad} = \underline{\quad}$ $\qquad 2 \times 3 \times 4 = \underline{\quad} \times 4 = \underline{\quad}$

5 $6 \times 9 \times 7 = 6 \times \underline{\quad} = \underline{\quad}$ $\qquad 6 \times 9 \times 7 = \underline{\quad} \times 7 = \underline{\quad}$

6 At a supermarket you have bought items costing 25p, 65p and 70p. Find the total cost in two ways, by adding different pairs first. Make sure you get the same result!

7 A box is a cuboid which is 4 cm high, 5 cm wide and 6 cm deep. Find its volume in two ways, by multiplying different pairs first. Make sure you get the same result!

So, if you have lots of numbers to be added or numbers to be multiplied, it doesn't matter which order you do them in. But what if you mix additions and multiplications? What about this expression?

$$3 + 5 \times 7$$

If you do the addition first, then you get

$$3 + 5 \times 7 = 8 \times 7 = 56$$

If you do the multiplication first, then you get

$$3 + 5 \times 7 = 3 + 35 = 38$$

These answers are different!

We need a rule for which operation is to be done first. Mathematicians have agreed that multiplication takes priority, so that when you have addition and multiplication in the same expression, the multiplication is done first.

So, in the expression $3 + 5 \times 7$, do the multiplication first.

$$3 + 5 \times 7 = 3 + 35 = 38$$

Type of calculator

If you have a scientific calculator, it obeys the rule of doing multiplication before addition. Try keying in the expression $3 + 5 \times 7$. You should get the answer 38. But, if you have an ordinary calculator, it may not give the correct answer.

Examples The entrance charge for a theme park is £10. You pay this and then go on six rides at £2 each. How much do you spend in total?

The total number of pounds is

$$10 + 6 \times 2 = 10 + 12 = 22$$

So you spend £22 in total.

Remember – multiply first!

A lorry is carrying three cars weighing 900 kg each, and four motorcycles weighing 150 kg each. What is the total weight that the lorry is carrying?

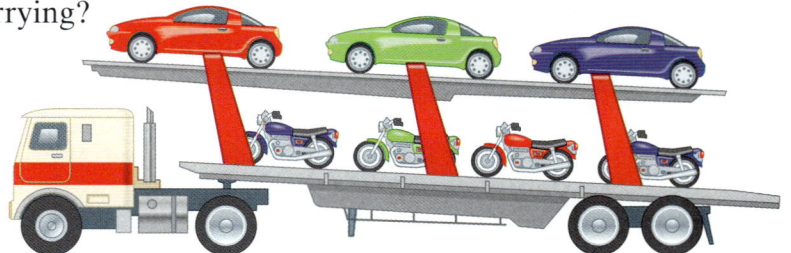

Multiply 900 by 3, and 150 by 4, and add. The expression is

$$3 \times 900 + 4 \times 150$$

The multiplications are done first.

$$3 \times 900 + 4 \times 150 = 2700 + 600 = 3300$$

So the total weight is 3300 kg.

Exercise 5.2

Evaluate the expressions in questions 1 to 6.

1 $2 + 3 \times 8$
2 $4 + 6 \times 3$
3 $3 \times 2 + 7$
4 $4 \times 8 + 3$
5 $3 \times 2 + 8 \times 4$
6 $5 \times 3 + 4 \times 7$

Remember –
multiply first!

7 What is the total cost of a cassette player costing £45 and six cassettes costing £4 each?
8 The crew of a rowing boat consists of a cox weighing 50 kg and eight oarsmen weighing 90 kg each. What is the total weight of the crew?
9 You buy three pizzas at £4 each and five colas at £1.20 each. What is the total cost?
10 In a house there are two electric fires switched on, each using 800 watts of power, and three electric lights, each using 100 watts. What is the total power consumption?

Examples If a pen costs 20p and a pencil costs 7p, then together they cost 27p. Suppose you bought three of each, what would the total cost be?

There are two ways you could work this out.
    You could think of it as three lots of 20p plus three lots of 7p.

$$3 \times 20 + 3 \times 7 = 81$$

The total cost is 81p.

Or you could think of it as three lots of 27p.

$$3 \times 27 = 81$$

The total cost is 81p.

Notice that the results are the same. Which way is easier?

Suppose a cup weighs 38 grams and a saucer weighs 22 grams. The total weight of cup and saucer set is 60 grams.
    What do six cup and saucer sets weigh?

Again there are two ways of doing it.

The total weight
is 360 grams.

Six lots of 38 grams plus six lots of 22 grams.

$$6 \times 38 + 6 \times 22 = 360$$

Six lots of 60 grams.

$$6 \times 60 = 360$$

The total weight
is 360 grams.

Again, the results are the same. You probably found the second way easier!

Exercise 5.3

A crayon costs 36p and an eraser costs 17p. Find the costs of the following.

1 **a** Two crayons and two erasers, bought separately.
   b Two sets of crayon and eraser.
2 **a** Three crayons and three erasers, bought separately.
   b Three sets of crayon and eraser.
3 **a** Four crayons and four erasers, bought separately.
   b Four sets of crayon and eraser.
4 If a cup costs £3.25 and a saucer costs £1.50, find in two different ways the total cost of six cups and saucers.
5 If a jar weighs 250 grams and its contents weigh 700 grams, find in two different ways the total weight of seven full jars.

Using brackets

When working out the pen and pencil problem on page 000, there were two ways of finding the total cost:

- multiply each price by 3, and then add
- add the prices, then multiply by 3.

The way of writing the first method is $3 \times 20 + 3 \times 7$. We handled expressions like this on page 70.

By convention, multiplication is done before addition. So if we want to do the addition first, we must alter the expression. The way of writing the second method involves **brackets**: $3 \times (20 + 7)$. We put brackets round the addition to show that it is done first.

For the cup and saucer problem on page 71, there were two ways of doing the calculation, depending on whether the addition was done at the end or the beginning.

$6 \times 38 + 6 \times 22$ doing the addition at the end
$6 \times (38 + 22)$ doing the addition at the beginning

Exercise 5.4

Write out these calculations in two ways, one doing addition at the end and the other doing addition at the beginning. Be sure to include the brackets for the second way. Check that your calculations give the same value.

1 Find the cost of 5 ballpoints and 5 pencils.
2 Find the cost of 3 pencils and 3 rulers.
3 Find the cost of 4 ballpoints and 4 erasers.
4 Find the cost of 8 pencils and 8 erasers.
5 Find the cost of 7 ballpoints and 7 rulers.
6 An electric fire uses 800 watts of power, and a light uses 60 watts. What is the total power consumption if three fires and three lights are switched on?
7 Jane has £150 and Joe has £200. If they change their money into Canadian dollars at 2 dollars per £, how much will they get?
8 The jacket of a suit weighs 12 ounces, and the trousers weigh 10 ounces. What is the total weight of five of these suits?
9 In the conversation at the beginning of this chapter, three people were discussing items costing 12 francs, 20 francs and 8 francs. These prices can be converted into £ by multiplying by 0.4. Show how to write the calculation, using brackets.

Use of calculator   If you have a scientific calculator, it will have buttons for brackets. The sequence for $3 \times (20 + 7)$ could be

$$\boxed{3}\ \boxed{\times}\ \boxed{(}\ \boxed{2}\ \boxed{0}\ \boxed{+}\ \boxed{7}\ \boxed{)}\ \boxed{=}$$

Practise the use of these buttons on some of the questions above. There is more guidance and practice on page 367.

Numbers on their own

These types of calculation don't have to be in terms of money or weight or power. The use of brackets also applies to numbers on their own.

$$3 \times (20 + 7) = 3 \times 20 + 3 \times 7$$
$$6 \times (38 + 22) = 6 \times 38 + 6 \times 22$$

Exercise 5.5

Write out these calculations as in the previous examples, checking that both ways of writing them give the same answer.

1 $9 \times (5 + 7)$ **3** $4 \times (5 + 3)$ **5** $15 \times (6 + 23)$
2 $8 \times (6 + 14)$ **4** $6 \times (4 + 13)$

We could describe the pattern in the answers to the last exercise in words.

Nine lots of 'five plus seven' is equal to nine lots of five plus nine lots of seven.

Eight lots of 'six plus fourteen' is equal to eight lots of six plus eight lots of fourteen.

Exercise 5.6

Describe the answers to the last three questions of exercise 5.5 in the same way.

Using algebra

Writing $9a + 9b$ means

'nine lots of a plus nine lots of b'

Using the pattern from exercise 5.3 this becomes

$9(a + b)$ (nine lots of 'a plus b')

In words

'nine lots of a plus nine lots of b is equal to nine lots of $a + b$'

To make it clear that the addition is done first, use brackets: write $(a + b)$. The rule is

$9(a + b) = 9a + 9b$

In the equation above, the left-hand side has brackets and the right-hand side doesn't. When we go from the left side to the right we are

expanding brackets

or

multiplying out brackets

or

opening out brackets.

Here is another way of looking at it. In the diagram below, there are two rectangles. One is 5 by r, the other is 5 by s. The first has area $5r$, the second has area $5s$.

total area $= 5r + 5s$

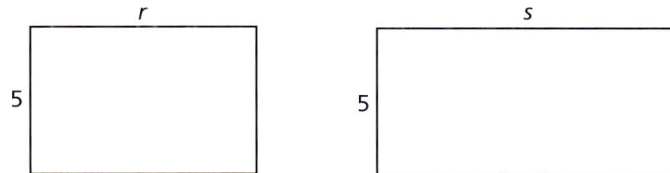

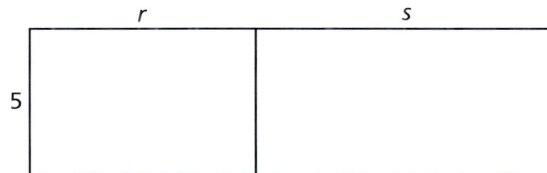

Now join them together as shown. There is now a long thin rectangle, of length $(r + s)$.

total area $= 5(r + s)$

These areas must be equal.

$$5r + 5s = 5(r + s)$$

Notice that the mathematics has become more abstract! We started with 5 pencils, then we went on to the number 5 by itself, now we go on to a letter by itself! It is all part of mathematics, making one expression deal with many different cases. The letter can stand for 5, or for 7, or for $3\frac{3}{4}$, or any number at all.

Exercise 5.7

• • • • • • • • •

Expand these expressions involving brackets.

1 $5(x + y)$
2 $7(x + y)$
3 $15(x + y)$
4 $15(c + d)$
5 $7(c + d)$

6 $5(g + h)$
7 $9(g + h)$
8 $13(a + b)$
9 $20(f + g)$
10 $4(k + n)$

> **Watch out!**
> In all these expressions, be sure to multiply *both* the terms inside the brackets by the number outside. Don't just multiply the first number inside.
> $3(x + y) = 3x + 3y$. It isn't $3x + y$.

Negative values in brackets

The values inside the brackets could be negative. The method of expanding brackets is just the same.

Remember that $6 \times (-5)$ is equal to -30.
Also $6 \times (-x)$ is equal to $-6x$.

So

> **Remember:**
> $6 \times x$ is the same as $6x$.

$$6 \times (9 - 5) = 6 \times 9 - 6 \times 5 = 54 - 30 = 24$$

and

$$6 \times (y - x) = 6 \times y - 6 \times x = 6y - 6x$$

Exercise 5.8

Multiply out these expressions.

1 $3 \times (11 - 4)$	**4** $7(x - y)$	**7** $3(x - v)$	**9** $3(y - x)$
2 $3(5 - 7)$	**5** $3(p - w)$	**8** $5(x - m)$	**10** $6(w - q)$
3 $4(8 - 2)$	**6** $7(a - b)$		

More complicated expressions

What if there is a more complicated expression inside the brackets? That doesn't matter. We can still multiply out.
Here three terms are multiplied together.

$$3(5x) = 3 \times 5 \times x$$

> **Remember:**
> $15 \times x$ is the same as $15x$.

We can multiply just the first two together.

$$3 \times 5 \times x = 15 \times x = 15x$$

Exercise 5.9

Open out these expressions.

1 $2(13x)$	**4** $7(6w)$	**7** $3(-2p)$	**9** $-4(3y)$
2 $5(6a)$	**5** $23(2x)$	**8** $2(-23x)$	**10** $-8(5m)$
3 $4(12x)$	**6** $4(-3x)$		

See what happens when we multiply out a more complicated bracket expression, and one with a negative term.

Examples Multiply out this expression $3 \times (5x + 7y)$.

$$3 \times (5x + 7y) = 3 \times 5x + 3 \times 7y$$
$$= 15x + 21y$$

Multiply out this expression $4(3p - 5q)$.

$$4(3p - 5q) = 4 \times 3p - 4 \times 5q$$
$$= 12p - 20q$$

Exercise 5.10

Multiply out these expressions.

1 $3(2s + 5t)$ **4** $6(2l + 4h)$ **7** $3(2a - 5b)$ **9** $8(3y - 5z)$
2 $5(3u + 2v)$ **5** $4(5c + 3d)$ **8** $4(6w - 2x)$ **10** $2(3a - 4b)$
3 $2(5r + 3z)$ **6** $5(2f + 6h)$

There might be both numbers and letters inside the brackets, as in this expression.

$$3(2x + 4)$$

Multiply out, just as you have done before.

$$3(2x + 4) = 3 \times 2x + 3 \times 4$$
$$= 6x + 12$$

Exercise 5.11

Expand these expressions.

1 $6(a + 5)$ **4** $3(x + 1)$ **7** $3(p - 1)$ **9** $3(4m + 8)$
2 $3(y + 7)$ **5** $4(y - 2)$ **8** $4(2x + 3)$ **10** $2(3k - 7)$
3 $2(w + 2)$ **6** $5(z - 4)$

Collecting like terms

Suppose Wayne has four oranges and three bananas, and that Jayne has four oranges and six bananas. Between them they have eight oranges and nine bananas.

There's no such thing as a 'banange', or an 'orana'!

That is as far as we can go. We can't combine the oranges and the bananas.

In algebra we like to make things as simple as possible. That's why so many questions ask you to 'simplify this expression'. But you mustn't simplify too much.

If you add $3x$ and $2x$, the result is $5x$. But you can't simplify $3x$ added to $2y$.

$3x + 2x = 5x$.
$3x + 2y$ cannot be simplified.
(This is like the bananas and oranges.)

When we simplify by writing $3x + 2x$ as $5x$, we are **collecting like terms**. The terms are 'like' as they both have xs in them. The expression $3x + 2y$ does not have like terms, because the x and y are different.

So you can always collect the same letters into a shorter expression.

$4a + 2a + 3a = 9a$ (just by counting)
$a + b + 2c + 3d$ cannot be simplified at all, as all the letters are different.

Example Simplify this expression.

$a + 3b + 2a + 4b$

We can collect the as, to obtain $3a$. We can collect the bs, to obtain $7b$.

$$a + 3b + 2a + 4b = a + 2a + 3b + 4b$$
$$= 3a + 7b$$

This now consists of unlike terms, so there is no further simplification.

Exercise 5.12

Simplify these expressions by collecting like terms.

1 $3x + 5x$ **4** $3z + 4z + 2z$ **7** $10x + 2x - 3x$ **9** $7z - z - 2z$
2 $7y + 2y$ **5** $6m - 2m$ **8** $5y - 2y + 3y$ **10** $8m - 5m - 2m$
3 $x + 2x + x$ **6** $4k - k$

Exercise 5.13

Simplify these expressions as far as possible by collecting like terms. If there is no possible simplification, write 'no simpler expression'.

1 $3m + 2n + 2m + 3n$ **6** $2p + 3q + 4r$
2 $4x + 2y + 3x + 8y$ **7** $2x + 3y + x$
3 $5q + 6p - 2q - p$ **8** $7p + 4q + r - q$
4 $9j + 2k - 3j - 5k$ **9** $4x + 3y - 3x - 2y$
5 $x + y + z$ **10** $4p - 5q - 3p + 4q$

Putting it all together

In this chapter we have introduced two main ideas: multiplying out brackets and collecting like terms. Now we put it all together. If we have two or more sets of brackets, we will need to use both techniques.

Example Simplify this expression.

$$2(3p + q) + 3(p + 2q)$$

Expand both sets of brackets.

$$2(3p + q) + 3(p + 2q) = 6p + 2q + 3p + 6q$$

Now collect like terms.

$$6p + 2q + 3p + 6q = 6p + 3p + 2q + 6q$$

So

$$2(3p + q) + 3(p + 2q) = 9p + 8q$$

This example involves negative terms.

Example Expand and simplify $3(m - 2n) + 4(3m - n)$ as far as possible.

$$3(m - 2n) + 4(3m - n) = 3m - 6n + 12m - 4n$$
$$= 3m + 12m - 6n - 4n$$
$$= 15m - 10n$$

Exercise 5.14

Expand these expressions, simplifying as far as possible

1 $2(x + 3y) + 3(2x + y)$
2 $4(p + 3q) + 6(2p + 3q)$
3 $5(2a + 3b) + 8(2a + 5b)$
4 $2(m + 4n) + 3(2m - n)$
5 $5(2x - 3y) + 7(3x - 2y)$

6 $3(c - 3d) + 4(2c - d)$
7 $3(2x + 7) + 2(3x + 2)$
8 $4(2y + 3) + 5(3y + 1)$
9 $2(2p + 4) + 3(3p - 5)$
10 $5(3q - 8) + 7(2q - 6)$

SUMMARY

- If an expression involves addition and multiplication, do the multiplication first.
- Two lots of 'three plus four' is the same as two lots of three plus two lots of four.
- Three lots of $(x + y)$ is equal to three lots of x plus three lots of y. In algebra this is written $3(x + y) = 3x + 3y$. Don't forget the brackets.
- Going from left to right in the equation above is **opening out brackets** or **expanding brackets** or **multiplying out brackets**. You must multiply *both* the terms inside the brackets by the term outside.
- You can sometimes simplify expressions by collecting occurrences of the same letter. This is **collecting like terms**.

Exercise 5A

1 Evaluate $3 + 6 \times 7$.
2 What is the total cost of a jacket at £80 and three shirts at £15 each?
3 A shirt costs £15 and a tie costs £5. What is the total cost of seven sets of shirt and tie? Work this out in two ways.
4 Expand $5(u + v)$.
5 Expand $6(x + 5)$.
6 Expand $8(2x + 3y)$.
7 Expand $2(3z - 5w)$.
8 Simplify as far as possible $13p + 7q - 3p - 2q$.
9 Expand and simplify $4(3x + y) + 5(2x - 2y)$.
10 Insert brackets in the following to make them true.
   a $3 \times 5 + 7 = 36$ **b** $2 + 3 \times 2 = 10$ **c** $8 - 3 \times 5 = 25$

Exercise 5B

1 Evaluate $4 \times 8 + 5 \times 2$.
2 A chair weighs 3 kg and a table weighs 10 kg. What is the total weight of five chairs and seven tables?

3 There are 200 ¥ (Japanese Yen) to the £. I have eight £10 notes and 12 £20 notes. How much do I get if I convert them all into ¥? Work this out in two different ways, and check that your answers are the same.

4 Expand $7(f - g)$.

5 Expand $7(p - 7)$.

6 Expand $3(6r + 2t)$.

7 Expand $5(4k - 25)$.

8 Simplify as far as possible $7x + 6x + y - 3x$.

9 Expand and simplify $8(2a - 5b) - 3(3a - 3b)$.

10 Insert brackets in the following to make them true.

  a $4 \times 3 + 2 \times 5 = 100$ **b** $2 \times 8 - 3 \times 4 = 40$ **c** $8 - 7 - 5 = 6$

Exercise 5C   Ma1

You have been expanding brackets by multiplying all the terms inside by the single term outside. What if there is more than one term outside? How do you expand this?

  $(a + b)(c + d)$

The diagram below shows a rectangle with width $(a + b)$ and height $(c + d)$. Write down its area in two ways, as a single large rectangle or as four smaller rectangles.

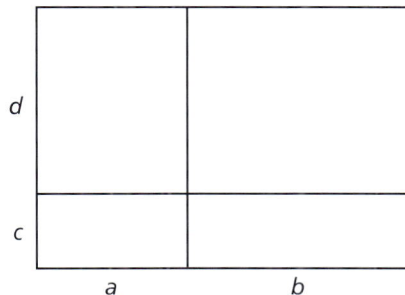

Have you found a rule for the expansion? Test it with some actual numbers for a, b, c and d.

Exercise 5D   Ma1

'Four 4s'

How many numbers can you make, using at most four 4s? You can combine them with $+$, $-$, $\times$ or $\div$, and you can use brackets if you want. You can put in a decimal point. Here are ways to make 1, 2 and 3, using only four 4s (or even fewer!)

  $1 = 4 \div 4$ $2 = 4 \div 4 + 4 \div 4$ $3 = 4 - 4 \div 4$

How many other numbers up to 20 can you get? Can you go beyond 20?

6 Coordinates

You can describe the positions of points by **coordinates**. Do you remember all the terms?

- Each axis is numbered uniformly, that is, with equal lengths between divisions.
- The numbering starts from 0 at the origin.

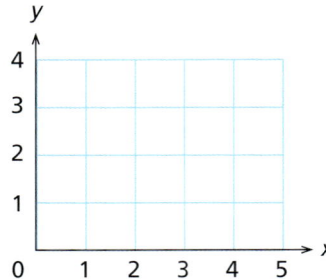

- The x-axis is numbered from left to right, and the y-axis is numbered from bottom to top. There are arrowheads at the ends of the axes to show that the numbers are increasing.

The position of a point is given by the values of x and y at the point. These are the **x-coordinate** and the **y-coordinate**. Always give the x-coordinate first.

In the diagram below, point A is at $x = 3$ and $y = 2$. These are the coordinates. We say that A is at $(3, 2)$.

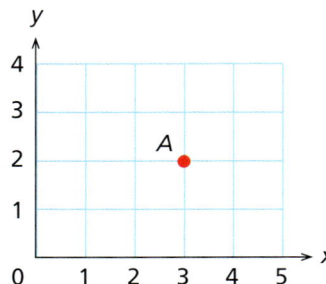

x comes before y in the alphabet.

Example Plot B at $(2, 1)$.

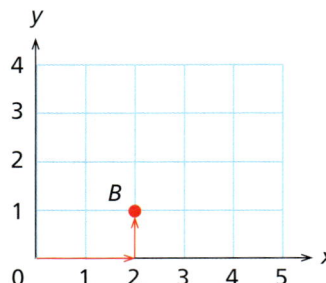

The x-coordinate is 2. Go 2 units along the x-axis, then up 1. Plot B as shown.

Exercise 6.1

(Mainly revision)

For each of questions 1 to 5 write down the coordinates of the point on the diagram.

1 *A*

2 *B*

3 *C*

4 *D*

5 *E*

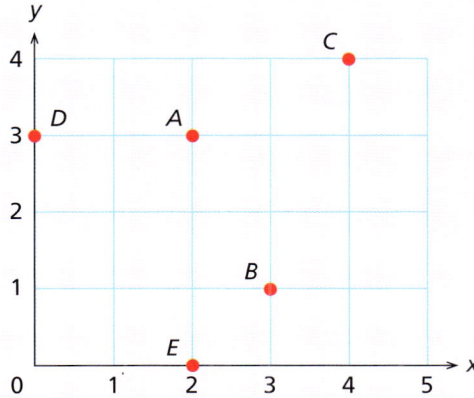

For questions 6 to 10 make a copy of the diagram below, and plot points with the given coordinates.

6 *F*, at $(1, 3)$

7 *G*, at $(4, 2)$

8 *H*, at $(3, 2)$

9 *I*, at $(3, 0)$

10 *J*, at $(0, 4)$

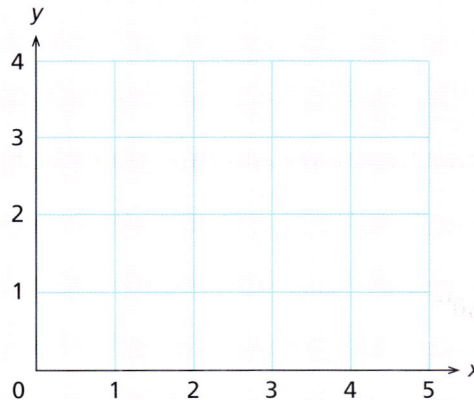

You will need:

• squared paper or graph paper
• ruler
• coloured pencils

throughout the work in this chapter.

Negative coordinates

Look at the coordinate grid for questions 6 to 10 above. What if a point lies underneath the *x*-axis? What if it lies to the left of the *y*-axis? We need to extend the grid so that it covers the whole region, not just the bit enclosed by the axes.

We use *negative* numbers for this. The position of any point on the line is given by a number. If the point is to the right of 0, the number is positive. If the point is to the left of 0, then the number is negative.

Remember:

Number lines

We can also draw a vertical number line (rather like the scale on a thermometer).

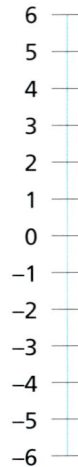

```
6 ─
5 ─
4 ─
3 ─
2 ─
1 ─
0 ─
−1 ─
−2 ─
−3 ─
−4 ─
−5 ─
−6 ─
```

Draw these number lines together on squared paper, so that their 0s coincide, as in the grid diagram below. Now any point on the grid can be represented by numbers on these lines. The sign of the numbers tells us whether the point is above or below the *x*-axis, and whether it is to the right or left of the *y*-axis.

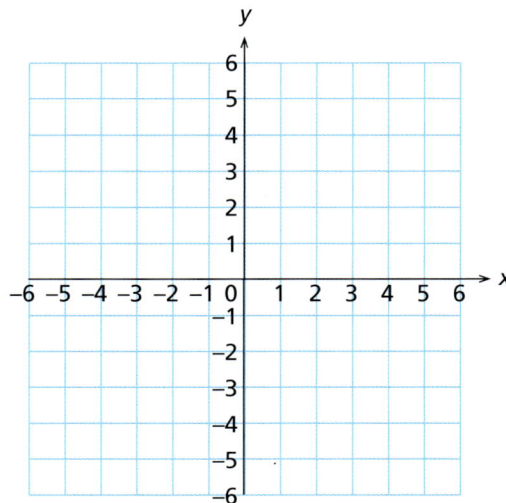

● *Points below the x-axis have negative y-coordinates.*
● *Points to the left of the y-axis have negative x-coordinates.*

Example The diagram below shows several points on a **coordinate grid**.

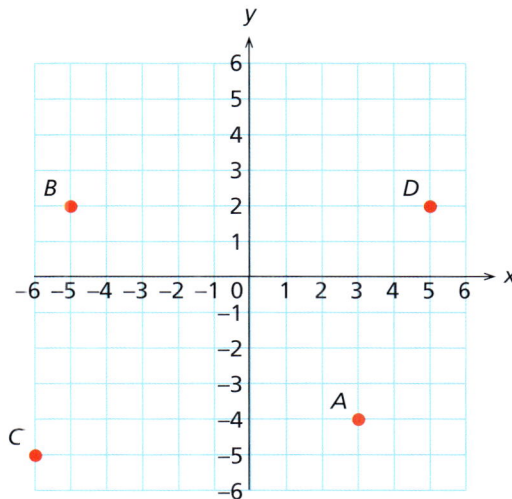

Point *A* is at (3, −4).   (it is below the *x*-axis)
Point *B* is at (−5, 2).   (it is left of the *y*-axis)
Point *C* is at (−6, −5). (it is left of the *y*-axis and below the *x*-axis)
Point *D* is at (5, 2).   (it is right of the *y*-axis and above the *x*-axis)

We use negative numbers in many practical situations. When it is below freezing outside, the temperature in degrees Celsius is negative! We can use negative coordinates to show negative temperatures.

Example Harry measured the temperature on several cold days. On day 1 it was −3°, on day 2 it was −2°, on day 3 it was 1° and on day 4 it was 4°. Plot these points on the coordinate grid below.

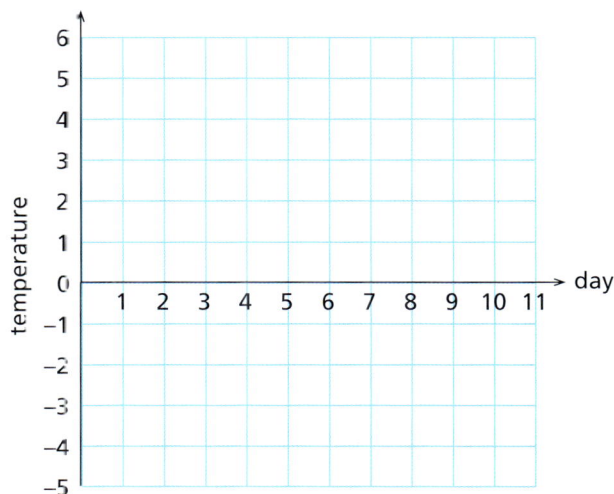

For day 1 on the *x*-axis, go down to −3 on the *y*-axis and mark a point. For day 2 go down to −2, for day 3 go up to +1 and for day 4 go to +4. The result is shown below.

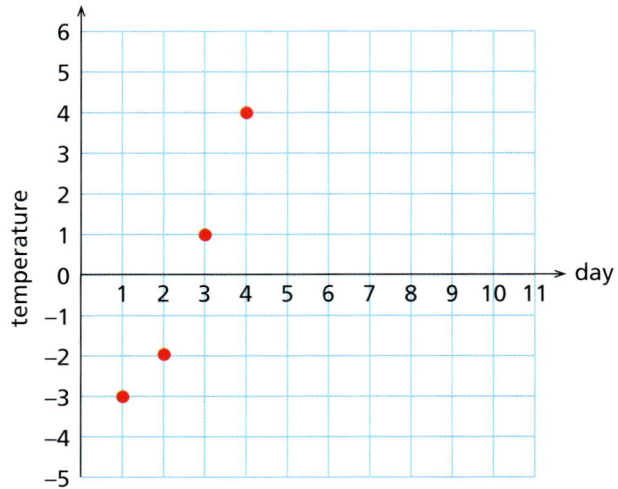

Exercise 6.2

For questions 1 to 7 write down the coordinates of the points on the diagram.

1 *A*

2 *B*

3 *C*

4 *D*

5 *E*

6 *F*

7 *G*

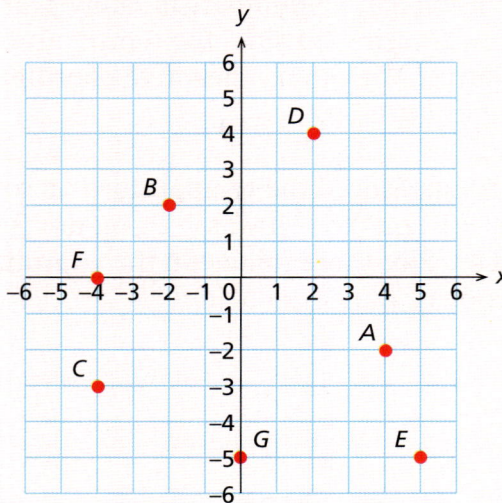

8 The level of a reservoir is measured above or below its normal level. The table below gives the level over a period of five days. Plot the values on a copy of the graph at the top of page 87.

day	1	2	3	4	5
change in level (inches)	1	0	−2	−3	−1

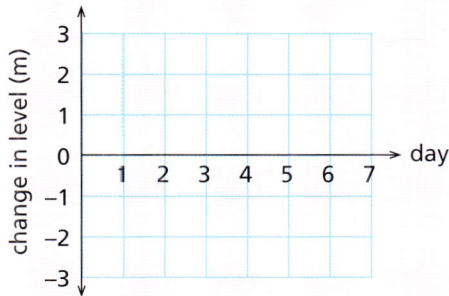

9 Rory is climbing up a mountain in Scotland. As he gets higher, the temperature falls. The table below gives the temperature at different heights. Plot the values on a copy of the graph below.

height (1000s of feet)	0	1	2	3
temperature (°C)	3	0	−1	−4

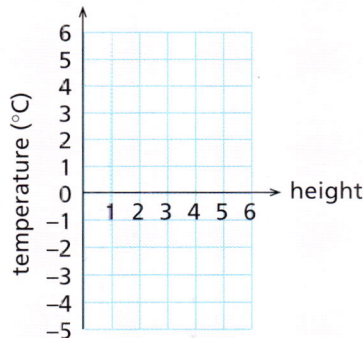

10 Some places such as Amsterdam in the Netherlands, are below sea level. So their height above sea level is negative. A railway train is travelling through the Netherlands, leaving at noon. The table below gives its height above sea level. Plot the values on a copy of the graph below.

time (p.m.)	1	2	3	4
height above sea level (m)	4	1	−4	3

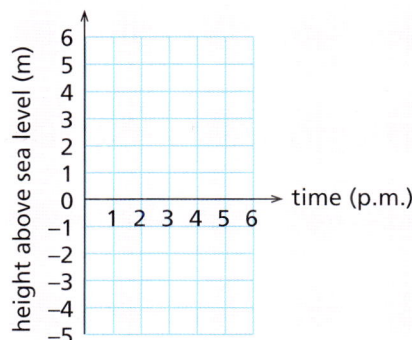

Exercise 6.3

Use squared paper to draw a grid, taking the origin near the middle of the paper. Mark out axes, and number both x-axis and y-axis from -6 to 6.
   Plot and label each of these points.

1 $A(-1, 2)$ **2** $B(2, -1)$ **3** $C(-2, -1)$ **4** $D(3, -4)$ **5** $E(-3, 4)$

Exercise 6.4

For these questions, set up a grid with the origin near the middle of the page, and both axes numbered from -8 to 8. Plot the points given, join them up and state the name of the shape.

1 $A(-2, 2)$, $B(-2, -2)$, $C(2, -2)$ and $D(2, 2)$
2 $(-3, 2)$, $(1, 2)$, $(1, 7)$ and $(-3, 7)$
3 $(-4, -5)$, $(1, -2)$, $(2, 2)$ and $(-3, -1)$
4 $(4, -2)$, $(7, -3)$, $(6, 0)$ and $(3, 1)$
5 $(-8, 5)$, $(-3, 6)$ and $(-6, 8)$
6 $(-3, 4)$, $(0, 0)$, $(5, 0)$ and $(2, 4)$
7 $(-6, 2)$, $(-1, 2)$, $(3, -1)$ and $(2, -4)$
8 $(-4, -6)$, $(3, -5)$ and $(-3, -3)$
9 $(-4, 4)$, $(-1, 3)$, $(0, 8)$ and $(-5, 7)$
10 $(-2, -5)$, $(3, -7)$, $(3, 3)$ and $(-2, 1)$

Exercise 6.5

Invent three shapes of your own. Give the coordinates to a partner, and see if he or she can draw the shapes on a grid like the previous one. Be sure to use both positive and negative numbers!

Exercise 6.6

Prepare a coordinate grid with the x-axis going from -15 to 15, and the y-axis going from -5 to 12.

1 Plot the points $A(-3, 5)$, $B(4, 5)$ and $C(4, 8)$. Find the coordinates of the point D such that $ABCD$ is a rectangle.
2 Plot the points $E(-13, 8)$, $F(-11, 5)$ and $G(-9, 8)$. Find the coordinates of the point H such that $EFGH$ is a rhombus.
3 Draw a line IJ with I at $(10, 0)$ and J at $(14, 3)$. JIK is an isosceles triangle with the x-axis as its line of symmetry. Find the coordinates of K.
4 Plot points at $L(-12, 1)$, $M(-12, -3)$ and $N(-8, 1)$. $MLNP$ is a square. Find the coordinates of P.
5 $QRST$ is a parallelogram with Q at $(-5, -3)$, R at $(3, -4)$ and S at $(6, 0)$. Find the coordinates of T.

Fractional coordinates

The points so far have all been on the gridlines. Their coordinates were always whole numbers.

Points aren't always so considerate! A point which falls between gridlines will have **fractional coordinates**. You need to be very careful when reading and writing down fractional coordinates.

Reading fractional coordinates

Examples Find the coordinates of points *A* and *B* on the diagram below.

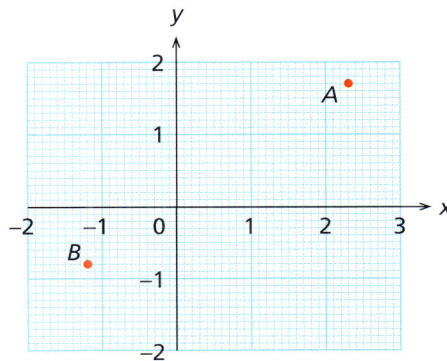

Count carefully the number of divisions between the whole numbers!

There are 10 small divisions (shown by thinner gridlines) between the whole numbers. So each division represents 0.1. This diagram shows a 'blown-up' picture of the square containing *A*.

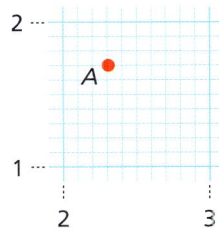

A is 3 small divisions to the right of 2 on the *x*-axis. Hence its *x*-coordinate is 2.3.

A is 7 small divisions above 1 on the *y*-axis. Hence its *y*-coordinate is 1.7.

So *A* is at (2.3, 1.7).

This diagram shows a blown-up picture of the square containing *B*. This is 2 small divisions to the left of −1 on the *x*-axis. Hence its *x*-coordinate is −1.2.

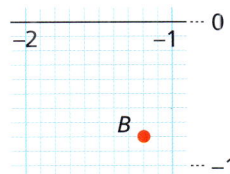

B is 8 small divisions below 0 on the *y*-axis. Hence its
y-coordinate is −0.8.

The diagram below gives the weights and diameters of coins. What
is the weight and diameter of the coin at point *A*?

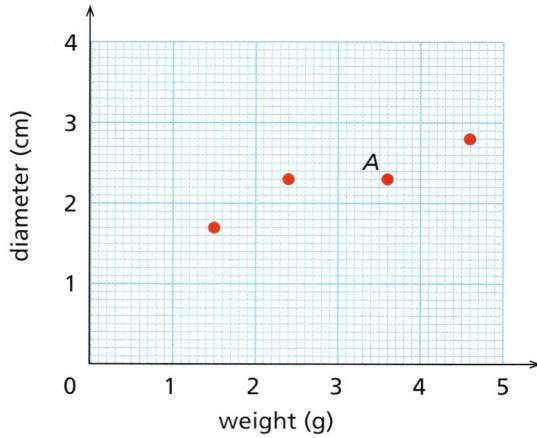

The point is 6 divisions to the right of 3.
  The weight of the coin is 3.6 grams.
  The point is 3 divisions above 2.
  The diameter of the coin is 2.3 cm.

Exercise 6.7

For questions 1 to 7 give the coordinates of the points.

1 *A*

2 *B*

3 *C*

4 *D*

5 *E*

6 *F*

7 *G*

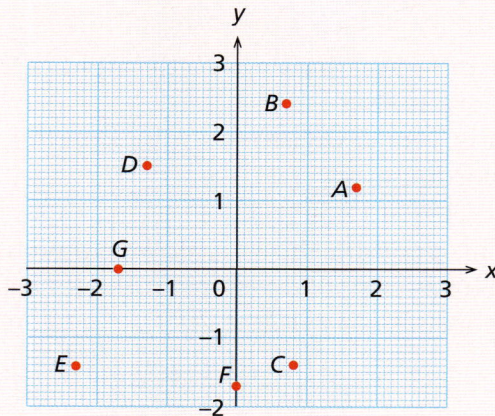

8 The diagram below describes a journey. It shows time in hours along the *x*-axis, and distance in miles up the *y*-axis. For the person at point *A*, find the time and the distance travelled.

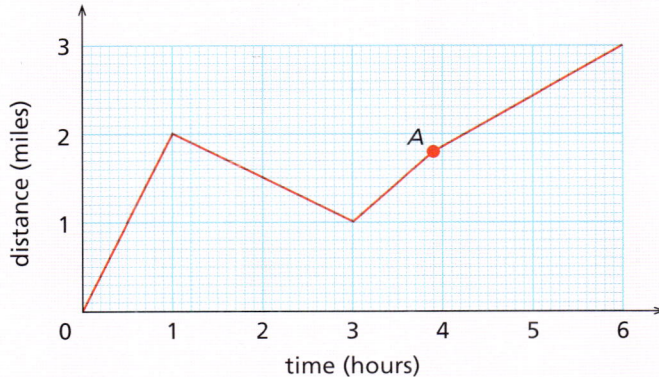

9 The diagram below is for conversion between currencies. It shows £ along the *x*-axis, and $ up the *y*-axis. Find the £ and $ conversion values at point *B*.

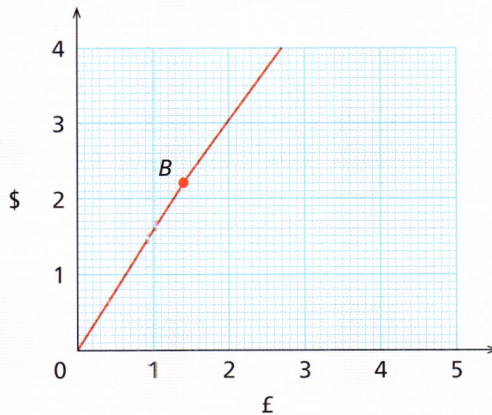

10 The diagram below shows change in temperature. It has time in hours along the *x*-axis, and temperature in °C up the *y*-axis. Find the time and temperature for point *C*.

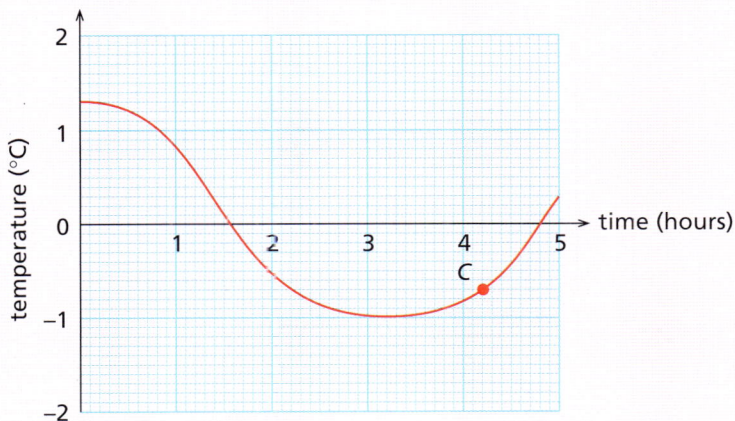

Plotting fractional coordinates

Example A coin has weight 2.9 grams and diameter 1.8 cm. Plot its point on the graph below.

> Be very careful when counting the divisions in fractional coordinates!

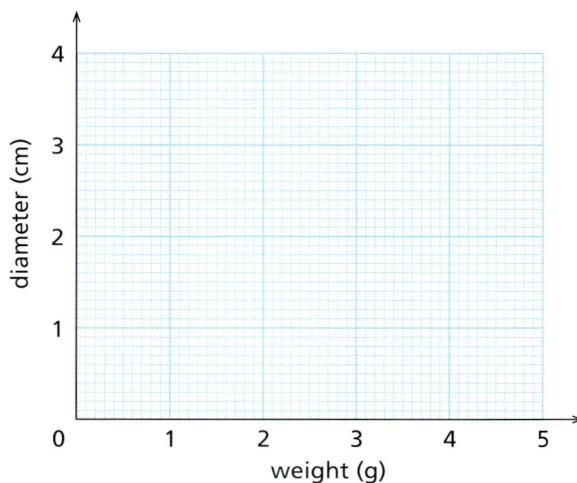

For the *x*-coordinate, go 9 divisions to the right of 2 (or 1 division to the left of 3).

For the *y*-coordinate, go 8 divisions above 1 (or 2 divisions below 2).

The result is as shown.

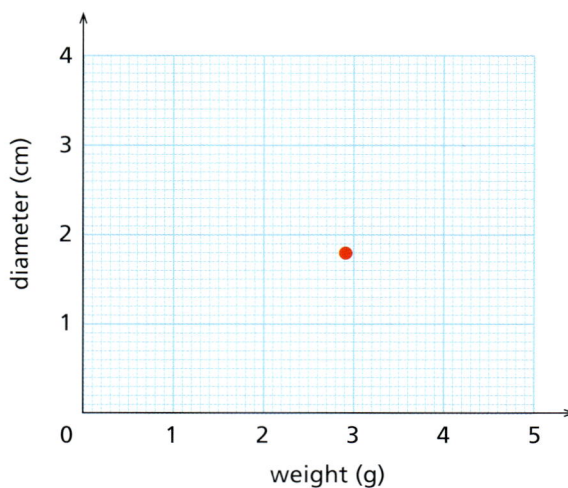

Exercise 6.8

For questions 1 to 5 make a copy of the grid and plot the given points on it.

1 A at (1.2, 2.6)

2 B at (2.5, 1.8)

3 C at (0.9, 2.4)

4 D at (−2.4, 1.3)

5 E at (−0.5, −1.7)

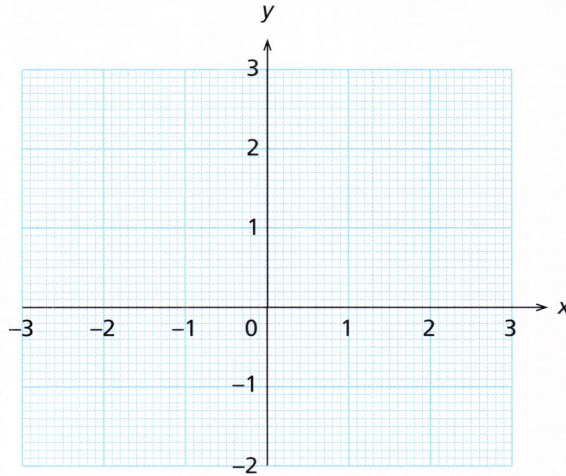

6 Make a copy of the grid below. Find the coins shown below. Measure their diameters and weigh the coins. Plot these measurements on the graph.

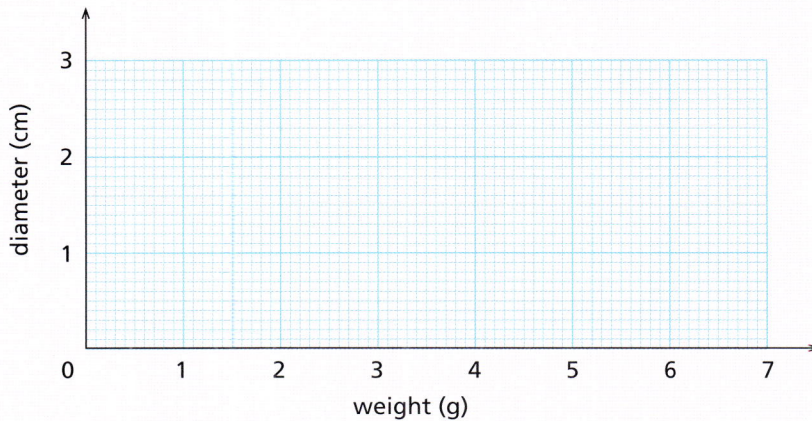

Smaller scales

In all our grids so far, the axes have been numbered in steps of 1. If we want to use larger numbers, we might need to number the axes in steps of 10. The grid might look like this.

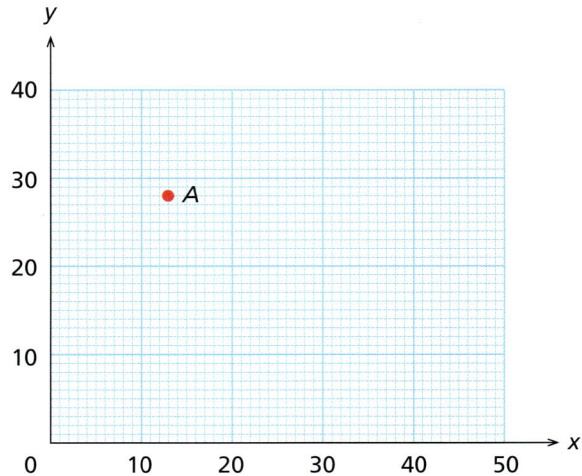

Example Find the coordinates of point *A* on the grid above. Plot the point *B* at (21, 18).

Point *A* is at 3 divisions to the right of 10. So the *x*-coordinate of *A* is 13.

Point *A* is at 7 divisions above 20. So the *y*-coordinate of *A* is 28. *A* is at (13, 28).

To plot *B*, go 1 division to the right of 20, and 8 divisions above 10. The result is shown below.

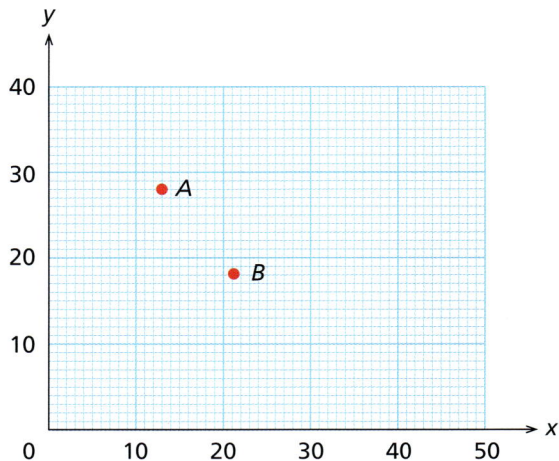

Exercise 6.9

For questions 1 to 5 write down the coordinates of the points on the diagram below.

1 *A*

2 *B*

3 *C*

4 *D*

5 *E*

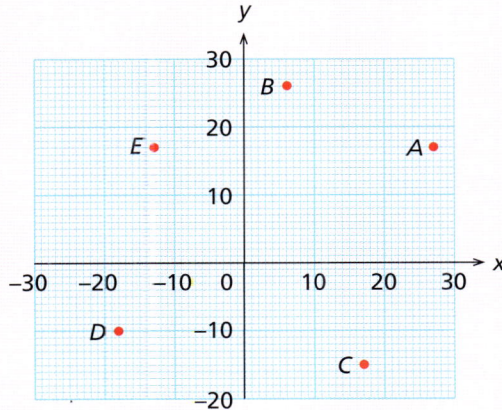

For questions 6 to 8 make a copy of the diagram below, and plot points with these coordinates.

6 $(12, 32)$

7 $(25, 22)$

8 $(7, 14)$

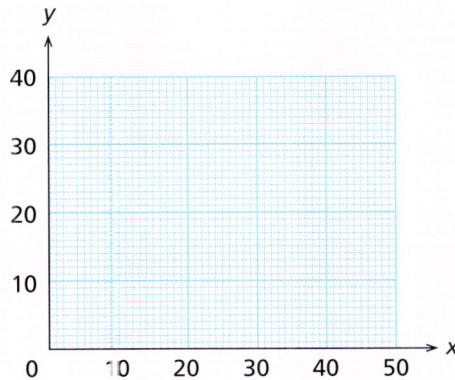

9 The diagram below shows marks in maths and physics tests. What are the marks for Martha, the person at point *A*?

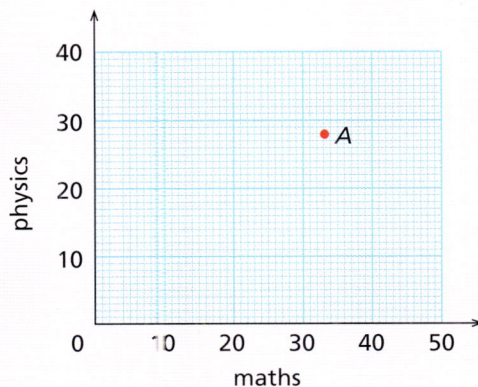

Make a copy of the diagram on page 95, and plot points for the marks of the following people.

	Eileen	Francis	Ganesh	Henry	Ivan
maths mark	43	48	39	36	44
physics mark	26	40	36	18	33

10 The temperatures (in °C) at Inverness (I) and Brighton (B) were measured at midnight. The diagram below shows the results. What were the temperatures at Inverness and Brighton on 1 October?

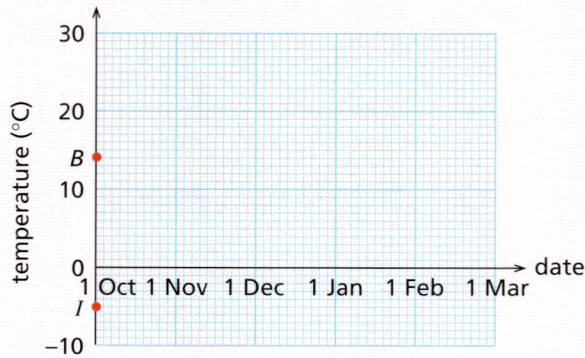

Make a copy of the diagram and on it mark the temperatures for the following dates.

	1 November	1 December	1 January	1 February	1 March
Inverness temp. (°C)	12	−1	−13	−8	17
Brighton temp. (°C)	18	9	−2	6	21

For even larger measurements, the axes may be numbered in steps of 100. In the diagram below, the gap between the main gridlines represents 100 m. Each small division represents 10 m.

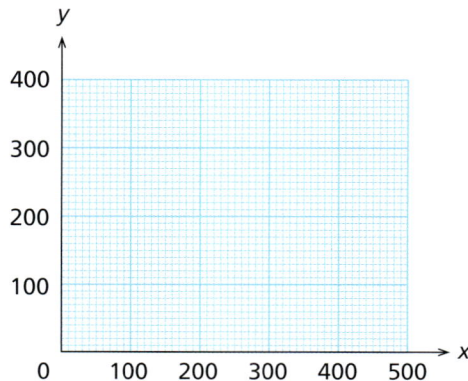

Exercise 6.10

For questions 1 to 3 write down the coordinates of the points on the diagram below.

1 *A*

2 *B*

3 *C*

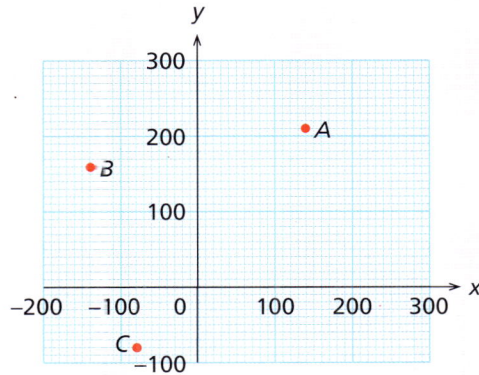

For questions 4 to 6 make a copy of the diagram below and plot points with these coordinates

4 (120, 300)

5 (−180, 170)

6 (250, −130)

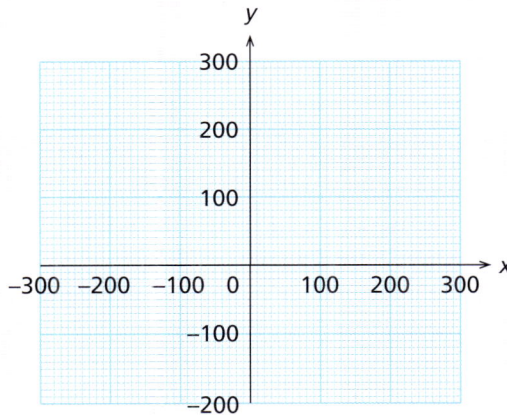

7 The diagram below is for measuring heights in centimetres and weights in pounds (lb). What height and weight are given for Andrew, at point *A*?

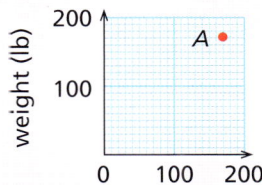

8 The heights and weights of three people are given in the table below. Make a copy of the diagram above, and plot the corresponding points on the graph.

	Samantha (S)	Theresa (T)	Ursula (U)
height (cm)	140	150	160
weight (lb)	110	130	140

Yet more complications! If there are 10 divisions per unit, then we can just count them. Each division represents 0.1. We have to be very careful if there are fewer divisions. Suppose there are 5 divisions per unit. Then each division represents $1 \div 5$, which is 0.2. It does not represent 0.1.

Example   Find the coordinates of the point A on the diagram below.

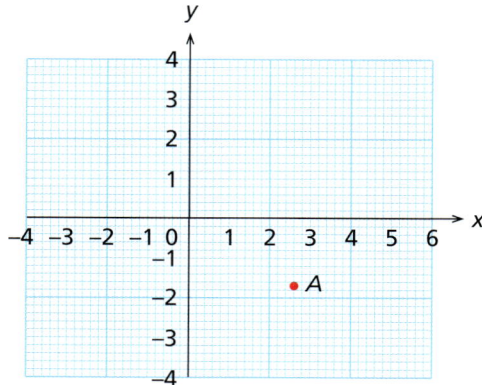

There are 5 divisions between units, hence each stands for 0.2.
Point A is 3 divisions to the right of 2. These divisions represent 3×0.2, that is 0.6. Hence its x-coordinate is $2 + 0.6$, or 2.6.
Point A is $3\frac{1}{2}$ divisions below -1. These divisions represent $3\frac{1}{2} \times 0.2$, that is 0.7. Hence its y-coordinate is $-1 - 0.7$, or -1.7.
The coordinates of A are $(2.6, -1.7)$.

Exercise 6.11

For questions 1 to 5 write down the coordinates of the points in the diagram below.

1 A

2 B

3 C

4 D

5 E

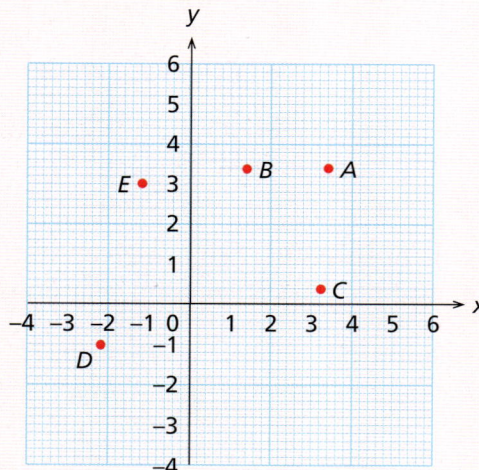

For questions 6 to 10 make a copy of the diagram below and plot the given points.

6 $(1.2, 1.8)$

7 $(2.3, 0.4)$

8 $(2.5, -2.3)$

9 $(-3.1, 1.4)$

10 $(0.7, 0.3)$

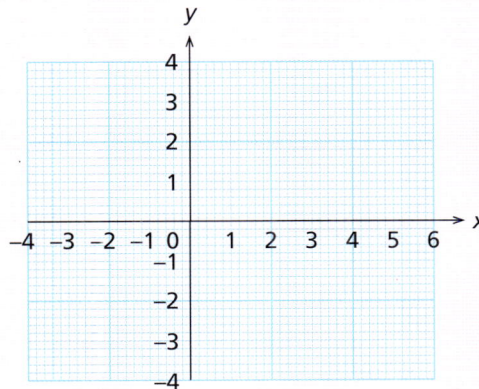

Lines on the grid

We can give the position of any point in a diagram, by giving its x- and y-coordinates. But what about the positions of *lines* in the diagram?

How could you describe the line shown in red in this diagram?

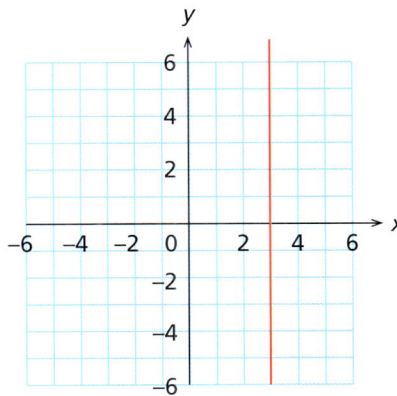

The line is vertical, and crosses the x-axis at $(3, 0)$.

This is rather long-winded! We need a shorter way to describe the line.

What points are on the line? Go up and down the line, and write down the coordinates of some points on it.

$$(3, 0)\ (3, 1)\ (3, 2)\ (3, 3)\ (3, 4)\ (3, -1)\ (3, -2)\ (3, -3)$$

What do you notice? It isn't hard! For each of the points, the x-coordinate is 3. For every point on this line, the x-coordinate is 3. So for each point, $x = 3$.

This identifies the line. The line has an equation, which is $x = 3$. Sometimes we go further, and say that the line **is** $x = 3$.

Exercise 6.12

Take the vertical line through (2, 0), shown in red on the diagram below. Write down the coordinates of several points on the line. What do they have in common? Write this down mathematically, by giving the **equation of the line**.

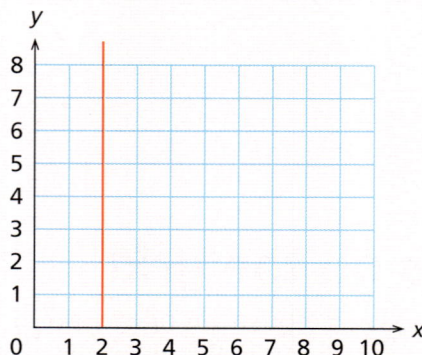

Exercise 6.13

Take the horizontal line through (0, 4), shown in dark blue on the diagram below. Write down the coordinates of several points on the line. What do they have in common? Write this down mathematically, by giving the equation of the line.

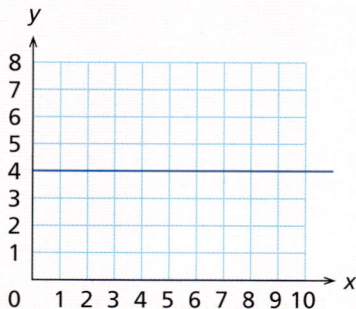

You should have found that the vertical line of exercise 6.12 has equation $x = 2$, and the horizontal line of exercise 6.13 has equation $y = 4$.

You don't have to go through all the business of writing down the coordinates of points.

- *Any vertical line has equation x = a number.*
- *Any horizontal line has equation y = a number.*

Example   Write down the equations of the blue and red lines shown on the diagram below.

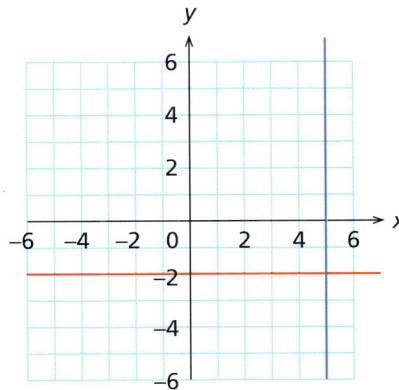

To find the number, see where the line crosses the axis!

The blue line is vertical, and crosses the *x*-axis at 5.
    The blue line has equation $x = 5$.
    The red line is horizontal, and crosses the *y*-axis at -2.
    The red line has equation $y = -2$.

Exercise 6.14

Write down the equations of the lines numbered 1 to 6 on the diagram below.

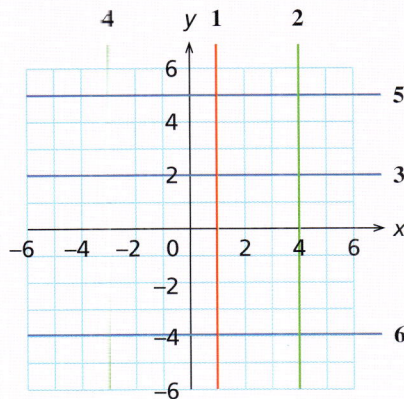

Make a copy of the diagram below, and draw some horizontal and vertical lines on it. Ask your neighbour to write down their equations.

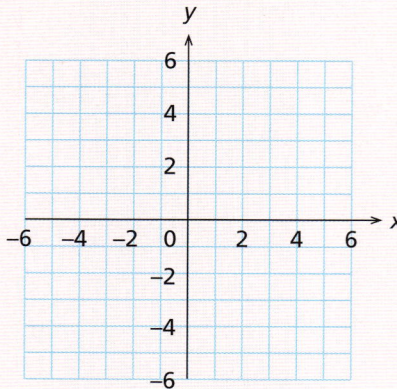

Exercise 6.15

There is one very important horizontal line: the *x*-axis itself! What is its equation?

Write down the coordinates of several points on the *x*-axis. What do they have in common? What is the equation of the *x*-axis?

Similarly, find the equation of the *y*-axis.

Be sure to get these the right way round!

You should find that the equation of the *x*-axis is $y = 0$, and that the equation of the *y*-axis is $x = 0$.

Diagonal lines

Look at the diagonal line through the origin, in the diagram below. $(1, 1)$ is a point on it. So is $(0, 0)$, and so on. Write down the coordinates of several points on this line. What do they have in common? Guess the equation of the line!

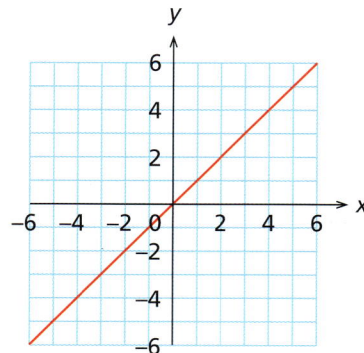

You should find that for all the points on this line, the *x*-coordinate is equal to the *y*-coordinate. For every point on the line, $y = x$. This is the equation of the line.

Exercise 6.16

Consider the other diagonal line through the origin, as shown in the diagram below. $(1, -1)$ is a point on it.

1 Write down the coordinates of some other points on the line.
2 What is the equation of the line?

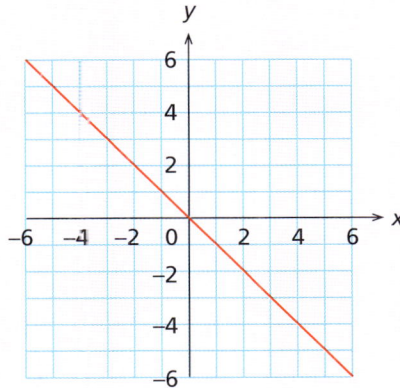

SUMMARY

- The position of a point on paper is given by two **coordinates**. The x-coordinate gives the horizontal distance from the **origin**. The y-coordinate gives the vertical distance from the origin. Always give the x-coordinate before the y-coordinate.
- If a point is to the left of the origin, its x-coordinate is negative. If a point is below the origin, its y-coordinate is negative.
- In some grids there are thinner gridlines, representing **fractional coordinates**.
- The scale of a grid does not have to be 1 unit per main gridline. It may be 10 units per gridline, or 100 units per gridline.
- You must be very careful when counting the divisions between main gridlines, particularly when each unit does not have 10 divisions.
- A vertical line has equation $x = $ a number. The number is where the line crosses the x-axis.
- A horizontal line has equation $y = $ a number. The number is where the line crosses the y-axis.
- The x- and y-axes have equations $y = 0$ and $x = 0$, respectively.
- The diagonal lines through the origin have equations $y = x$ and $y = -x$.

Exercise 6A

1 Find the coordinates of point *A* on the diagram below.

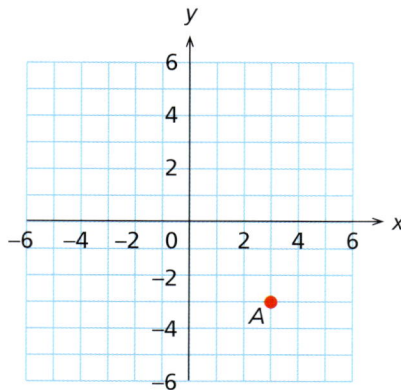

2 The diagram below shows part of a spreadsheet. What is the position of the shaded cell?

3 Draw coordinate axes, with *x* and *y* going from −5 to 5. Plot points at (1, 1), (−4, 1), (−1, 5) and (4, 5). Join up the points. What sort of shape do you have?

4 On your axes for question 3, plot points *A*(3, 2) and *B*(4, −1). The triangle *ABC* is isosceles, with a vertical line of symmetry. Where is *C*? (There are two possible answers!)

5 What are the coordinates of point *A* on the diagram below?

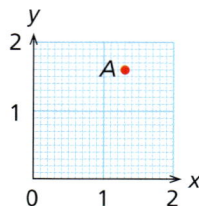

6 The diagram below shows prices of items bought either in a high street store or by mail order. What are the prices of the shirt at point *A*?

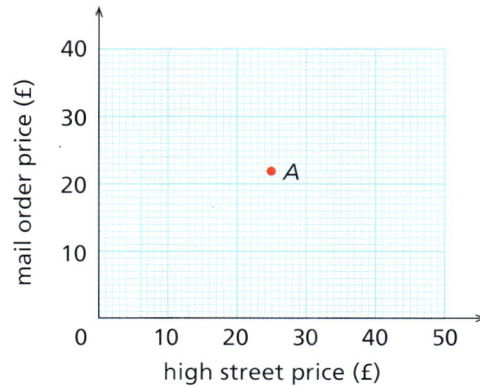

7 Make a copy of the diagram in question 6. On it mark the prices of the items in the table below.

	radio	pair of trainers	camera
high street price	£28	£31	£49
mail order price	£24	£29	£40

8 What are the coordinates of point *A* on the diagram below?

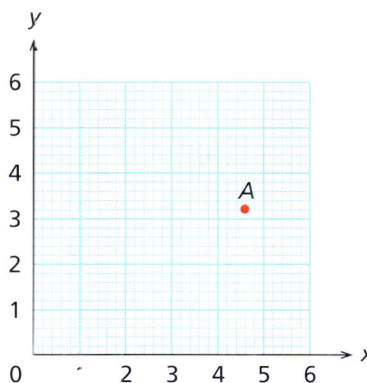

9 What are the equations of the lines shown in the diagram below?

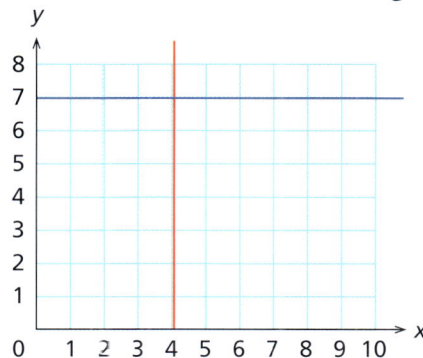

10 Mark out coordinate axes. Draw the lines with equations $x = 2$ and $y = 3$.

Exercise 6B

1 Find the coordinates of point X on the diagram below.

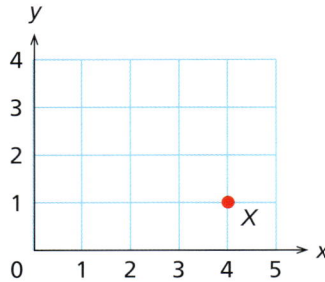

2 A triangle ABC has vertices at $A(2, 1)$ and at $B(0, 4)$. The triangle is symmetric about the y-axis. What are the coordinates of C?

3 The diagram below is to show the temperature over a 6 hour period. The table below gives the values. Plot points on a copy of the graph to show the temperatures.

time (hours)	0	1	2	3	4	5	6
temperature (°C)	1	2	0	−1	−3	−4	−2

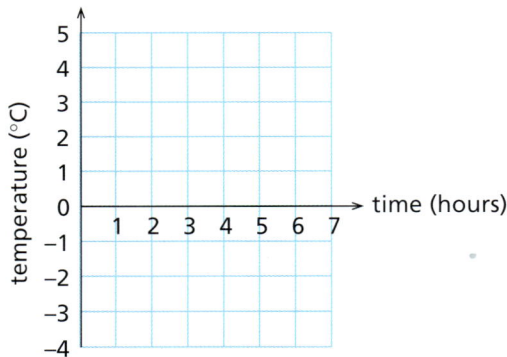

4 Draw coordinate axes, with x and y going from -6 to 6. Plot points at $(1, 1)$, $(5, 4)$ and $(-2, 5)$. What sort of triangle is this?

5 What are the coordinates of point Y on the diagram below?

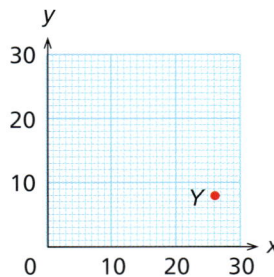

6 A maths exam has two parts: a written test and a project. The diagram below is for the marks. Anna's marks are at point *N*. What did she get?

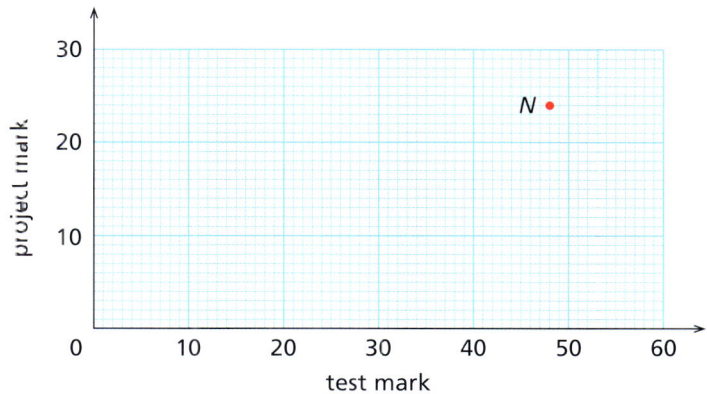

7 Make a copy of the diagram in question 6. On it plot points for the marks of the following people.

	George	Sidney	Shirley	Jacqui
project mark	10	27	16	17
test mark	36	58	44	53

8 What are the coordinates of point *Z* on the diagram below?

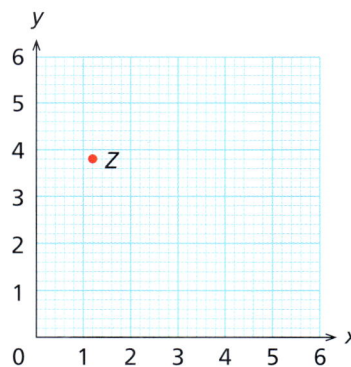

9 What are the equations of the lines shown in the diagram below?

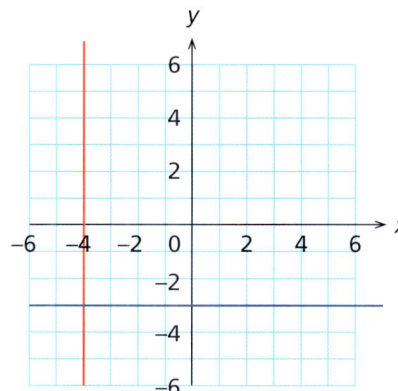

10 Mark out coordinate axes. Draw the lines with equations $y = x$ and $y = -x$.

Exercise 6C (Ma1)

Coordinates are useful! They can help you from getting lost in the country! On a map, a position is given by a **grid reference**. Get hold of an Ordnance Survey map (of the Landranger series). You will see a grid of blue lines on the map. The grid is numbered on the sides of the map, by two-digit figures.

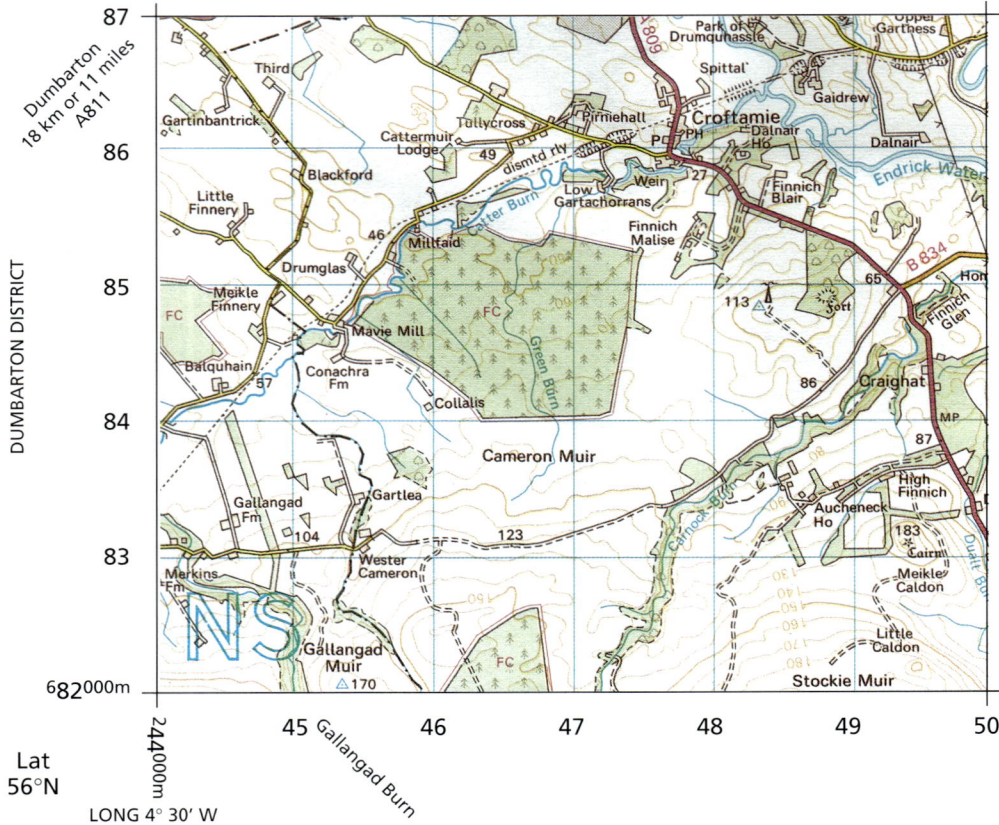

Give the East–West position first. Suppose a place is between 45 and 46 on the horizontal axis, between $\frac{3}{10}$ and $\frac{4}{10}$ of the distance between 45 and 46. Then its *easting* (equivalent to the *x*-coordinate) is 453.

Now give the North–South position. Suppose the place is between 84 and 85 on the vertical axis, between $\frac{4}{10}$ and $\frac{5}{10}$ of the distance between 84 and 85. Then its *northing* (equivalent to the *y*-coordinate) is 844.

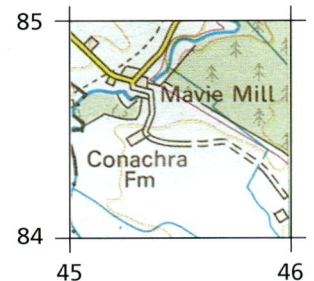

1 The grid reference of the place is 453844. What do you find at this grid reference?
2 Find the grid references of some other places on the map. Ask a partner to find these points. Then change roles!

Exercise 6D Ma1

You should always be careful to set out your axes correctly.
What is wrong with each of the graphs below?

1

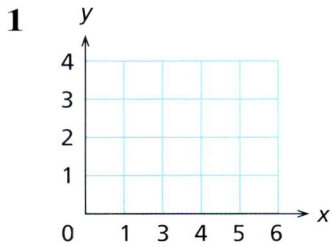

2

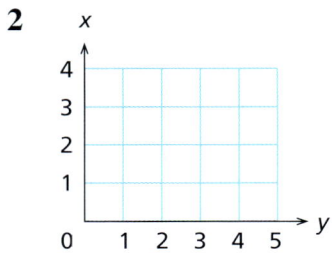

3

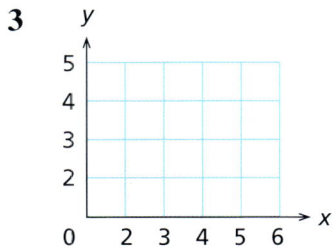

7 Decimal places and significant figures

The school running track is 400 metres long. Three boys make up a team to enter a relay race, in which they will go round the track once between them. If their stages are equal, how long is each stage?

> That's easy – just divide 400 by 3. That'll give the answer.

Using a calculator, Aidan finds that 400 divided by 3 gives 133.3333333.

> We can't measure a distance that accurately. Let's just measure out 133 m and forget those decimal numbers. Remember, we did it in maths; it's called 'rounding to the nearest metre'.

> Or, we could measure out 130 m. That's 'rounding to the nearest 10 m'.

They measure out 133 m, and so divide the track into three roughly equal stages. Then Aidan has an idea.

> We've been working in metres – what if we work in kilometres? Our running track is 0.4 km long, divide that by 3 and we get 0.13333333 km. Round that to the nearest km, that's 0 km! That can be my part of the race, I'll run 0 km!

> Me too – my third of the race is 0.13333333 km too, so if I round that to the nearest km, that's 0 km!

> Hang on! That's not fair, you've left the whole race to me!

What has gone wrong?
Before we can answer this, we need to revise and do more work on decimals and rounding.

Decimal places

Remember:
$0.5 = \frac{5}{10}$
$0.05 = \frac{5}{100}$

We can write parts of a whole number either as fractions or as decimals. The decimal point separates the whole number part from the fractional part. The digit immediately to the right of the decimal point is in the **first decimal place**, and represents the number of tenths. The digit to the right of that is in the **second decimal place**, and represents the number of hundredths, and so on. Decimal place is often written d.p. for short.

Consider 2.458.

UNITS	•	tenths	hundredths	thousandths
2	•	4	5	8
	•	1st d.p.	2nd d.p.	3rd d.p.

The digit in the first decimal place is 4. It represents 4 tenths, $\frac{4}{10}$.
The digit in the second decimal place is 5. It represents 5 hundredths, $\frac{5}{100}$.
The digit in the third decimal place is 8. It represents 8 thousandths, $\frac{8}{1000}$.
You must be careful with 0s. Consider 3.07. The 0 is in the first decimal place, and 7 is in the second decimal place. It represents $\frac{7}{100}$, not $\frac{7}{10}$.

UNITS		tenths	hundredths
3	•	0	7

Exercise 7.1

(Mainly revision)

1 For the number 5.123, what digit is in the first decimal place?
2 For the number 21.536, what digit is in the second decimal place?
3 For the number 0.04632, what digit is in the third decimal place?
4 For the number 3.2076, what digit is in the second decimal place?
5 Consider the number 2.463.
   a What fraction does the 4 represent?
   b What fraction does the 6 represent?
   c What fraction does the 3 represent?

6 Write the following as decimals:

a $3 + \frac{6}{10} + \frac{7}{100}$

b $7 + \frac{3}{10} + \frac{5}{100}$

c $2 + \frac{1}{10} + \frac{4}{100} + \frac{3}{1000}$

Rounding to the nearest whole number

Remember:

Rounding means the same as approximating.

In real life, you cannot always measure a length or a weight or a time exactly. Sometimes you have to give an approximate answer. We can round measurements to the nearest whole number, or to the nearest 10, or to the nearest 100.

In this picture the width of a photograph is being measured. The width is between 12 cm and 13 cm. We can see that it is closer to 13 cm, so that is what we give for its width. To the nearest whole number of centimetres, the width is 13 cm.

What if the measurement lies exactly half-way between two values? Look at this picture of a thermometer in which the temperature seems to be exactly half-way between 8° and 9°. If we are rounding to the nearest whole number of degrees (°), the rule is that we round up, in this case to 9°.

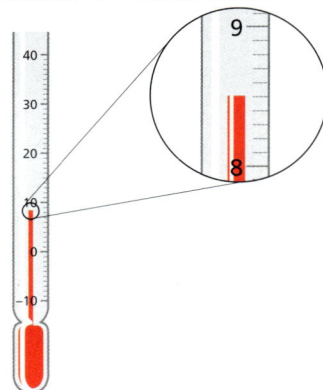

● *This is the rule for **rounding to the nearest whole number**. Look at the digit in the first decimal place. If it is 5 or more, round up. If it is 4 or less, round down. Note that the half-way point, of 5 in the first decimal place, is rounded **up**.*

Round 2.764 up to 3
Round 5.395 down to 5
Round 6.5 up to 7   (This is the rule.)

Exercise 7.2

In questions 1 to 5, round the numbers to the nearest whole number.

1 34.29
2 33.56
3 98.058
4 23.5
5 2.499

> To round to a whole number, look at the digit in the first decimal place to see whether to round up or down.

Evaluate the expressions in questions 6 to 10, giving your answers to the nearest whole number.

6 34.6 + 33.2
7 3.7 + 3.6
8 3.45 × 32.1
9 0.982 × 56
10 100 ÷ 7

> Do the calculation **before** the rounding.

Rounding to one decimal place

We have rounded numbers to the nearest whole number. If we need greater accuracy than that, we can round to the nearest tenth of a whole number, that is to the nearest 0.1. This is **rounding to one decimal place**.

Look at this picture of the measurement of a coin. It has been magnified, to show it more clearly. The width of the coin is between 1.7 cm and 1.8 cm. It is closer to 1.7 cm, so we round the answer down. The width is 1.7 cm, correct to one decimal place.

● *This is the rule for **rounding to one decimal place**. Look at the digit in the second decimal place. If this is 5 or more, round up. If it is 4 or less, round down.*

Round 1.682 up to 1.7
Round 3.931 down to 3.9
Round 4.65 up to 4.7   (This is the rule.)

Exercise 7.3

In questions 1 to 5, round the numbers to one decimal place.

> To round to **one** decimal place, look at the digit in the **second** decimal place.

1 34.69
2 1.353
3 2.35
4 0.17
5 0.43
6 The height of a building is 9.86 m. Round this to one decimal place.
7 The weight of a parcel is 4.23 kg. Round this to one decimal place.
8 An athlete's time to run a race is 28.43 seconds. Round this to one decimal place.
9 The temperature is measured at 22.45 °C. Round this to one decimal place.
10 The population of a city is 3.582 millions. Round this to one decimal place.

Rounding to a given number of decimal places

You have been rounding numbers to one decimal place. You can extend this, to round numbers to two decimal places, to three decimal places, and so on.

● *This is the rule for **rounding to two decimal places**. Look at the number in the third decimal place. If it is 5 or more, round up. If it is 4 or less, round down.*

Round 4.827 up to 4.83
Round 3.162 down to 3.16
Round 5.835 up to 5.84   (This is the rule.)

Rounding to a given number of decimal places • 115

Exercise 7.4

In questions 1 to 5, round the numbers to two decimal places.

1 23.452
2 1.987
3 0.255
4 35.847
5 0.028
6 The weight of a lorry is 4.533 tonnes. Round this to two decimal places.
7 The width of a table is 1.2934 m. Round this to two decimal places.
8 The weight of a letter is 43.328 g. Round this to two decimal places.
9 A bottle of juice contains 0.8435 litres. Round this to two decimal places.
10 A salary, in £1000s, is 34.426. Round this to two decimal places.

Exercise 7.5

See if you can write out the rule for rounding a number to three decimal places.

● *The rule is: look at the digit in the fourth decimal place, if this is 5 or greater, round up. If it is 4 or less, round down.*

Similar rules apply to rounding to four decimal places, and so on. These numbers have been rounded to three decimal places.

Round 3.4673 down to 3.467.
Round 0.0357 up to 0.036.
Round 0.1975 up to 0.198.

These numbers have been rounded to four decimal places.

Round 1.38473 down to 1.3847.
Round 23.11066 up to 23.1107.
Round 1.02845 up to 1.0285.

Exercise 7.6

Round the following to the given number of decimal places.

1 3.4786, correct to three decimal places.
2 5.6392, correct to three decimal places.
3 0.2642, correct to three decimal places.
4 25.29574, correct to four decimal places.

    5 0.001587, correct to four decimal places.
    6 2.583744, correct to five decimal places.
    7 0.3774581, correct to five decimal places.
    8 21.5330299, correct to five decimal places.
    9 0.000236, correct to four decimal places.
  10 0.00025976, correct to five decimal places.

Leaving in 0s

> Be careful! Remember that, to round to **three** decimal places, look at the digit in the **fourth** decimal place.

Sometimes, when you round a number to two decimal places, the digit in the second decimal place is 0. For example, consider 4.503 and 4.498.

For 4.503, the digit in the third decimal place is 3. Round down to 4.50.

For 4.498, the digit in the third decimal place is 8. When we round up, the 9 becomes 10. So the number is 4.50, to two decimal places.

> In other words, leave as 4.50, not as 4.5.

In both cases, you must **leave the 0 in the answer**.
  This shows that you have rounded to two decimal places, not to one decimal place. It shows that your answer is accurate to a hundredth.

Exercise 7.7

Round the following to the given number of decimal places.

    1 3.03, correct to one decimal place
    2 6.04, correct to one decimal place
    3 9.103, correct to two decimal places
    4 0.604, correct to two decimal places
    5 3.872048, correct to four decimal places
    6 4.96, correct to one decimal place
    7 23.96, correct to one decimal place
    8 0.4981, correct to two decimal places
    9 1.08497, correct to four decimal places
  10 0.4595, correct to three decimal places

Converting fractions to decimals

Converting some fractions to decimals is easy.

$$\tfrac{1}{2} = 0.5 \qquad \tfrac{1}{4} = 0.25 \qquad \tfrac{3}{8} = 0.375 \qquad \tfrac{2}{5} = 0.4$$

What about $\tfrac{1}{3}$? When you divide 1 by 3 on a calculator, the 3s fill up the screen.

> To convert a fraction to a decimal, divide the top number (numerator) by the bottom number (denominator).

$$0.333333333$$

Even this is only an approximation. To give $\tfrac{1}{3}$ exactly as a decimal, the 3s would have to go on for ever. This decimal is a **recurring** decimal.

$$\tfrac{1}{3} = 0.333\ldots$$

We show that the 3s go on for ever by putting a dot on top of the first 3.

$$\tfrac{1}{3} = 0.\dot{3}$$

Similarly

$$\tfrac{1}{9} = 0.111111\ldots \text{ or } 0.\dot{1} \qquad \text{(the 1s are repeated for ever)}$$
$$\tfrac{1}{6} = 0.16666\ldots \text{ or } 0.1\dot{6} \qquad \text{(the 6s are repeated for ever)}$$

The fraction $\tfrac{1}{11}$ is equal to 0.090909..., where the pattern 09 is repeated for ever. We put dots on top of the repeating pattern.

$$\tfrac{1}{11} = 0.\dot{0}\dot{9} \qquad \text{(the pattern 09 is repeated for ever)}$$

The fraction $\tfrac{1}{7}$ is more complicated. The repeated pattern has six digits. We put a dot on top of the first and the last digit of the pattern.

> $\tfrac{1}{3} = 0.\dot{3}$
>
> $\tfrac{1}{6} = 0.1\dot{6}$
>
> $\tfrac{1}{7} = 0.\dot{1}4285\dot{7}$
>
> $\tfrac{1}{9} = 0.\dot{1}$
>
> $\tfrac{1}{11} = 0.\dot{0}\dot{9}$

$$\tfrac{1}{7} = 0.142857142857\ldots = 0.\dot{1}4285\dot{7} \qquad \text{(the pattern 142857 is repeated for ever)}$$

Exercise 7.8

Use a calculator to write these fractions as decimals. If the decimal repeats, use the dot notation.

1 $\tfrac{3}{4}$

2 $\tfrac{5}{8}$

3 $\tfrac{3}{5}$

4 $\tfrac{2}{3}$

5 $\tfrac{2}{9}$

6 $\tfrac{8}{9}$

7 $\tfrac{4}{11}$

8 $\tfrac{8}{33}$

9 $\tfrac{3}{7}$

10 $\tfrac{4}{13}$

Exercise 7.9

Write out the first eight digits for each of the following decimal expansions.

1 $0.\dot{7}$

2 $0.\dot{8}$

3 $0.2\dot{8}$

4 $0.1\dot{3}$

5 $0.0\dot{3}$

6 $0.\dot{2}\dot{0}$

7 $0.\dot{1}2\dot{3}$

8 $0.50\dot{8}$

9 $0.\dot{6}0\dot{6}$

10 $0.\dot{1}25\dot{3}$

Often we don't need a fraction to be absolutely accurate. We can give its value to a certain number of decimal places.

The last 6 changes to a 7 (rounded up).

$\frac{1}{3} = 0.3333$, correct to four decimal places

$\frac{1}{6} = 0.1667$, correct to four decimal places

$\frac{1}{7} = 0.143$, correct to three decimal places

Exercise 7.10

Use a calculator to write each of the following fractions as a decimal, rounding its value to the given number of decimal places.

1 $\frac{2}{9}$ correct to three decimal places

2 $\frac{3}{11}$ correct to three decimal places

3 $\frac{2}{3}$ correct to four decimal places

4 $\frac{7}{33}$ correct to two decimal places

5 $\frac{5}{7}$ correct to four decimal places

6 $\frac{6}{13}$ correct to three decimal places

7 $\frac{8}{17}$ correct to four decimal places

8 $\frac{1}{99}$ correct to two decimal places

9 $\frac{1}{999}$ correct to two decimal places

10 $\frac{32}{33}$ correct to three decimal places

Rounding calculations

For many other calculations also, you may need to round your answer to a given number of decimal places.

● *Don't round at the beginning, or half-way through a calculation. Keep as much accuracy as you can until the very end.*

Example Evaluate 5.16×2.87, giving your answer correct to one decimal place.

Your calculator will give you

$$5.16 \times 2.87 = 14.8092$$

Rounding this to one decimal place, we get 14.8.

$$5.16 \times 2.87 = 14.8, \text{ correct to one decimal place}$$

Suppose you round the numbers at the beginning as well as at the end. You round 5.16 to 5.2, and you round 2.87 to 2.9.

$$5.2 \times 2.9 = 15.08$$
$$= 15.1, \text{ correct to one decimal place}$$

This answer is quite different! This example shows that you must do the rounding at the end.

Exercise 7.11

Use your calculator to evaluate the following, rounding your answer to the given number of decimal places.

1 4.62×9.19, correct to one decimal place
2 2.33×1.477, correct to two decimal places
3 $47 \div 35$, correct to two decimal places
4 $44 \div 123$, correct to four decimal places
5 $23.77 \times 1.22 \times 3.28$, correct to two decimal places
6 A motorist buys 55.3 litres of petrol, at £0.66 per litre. What is the total cost? Give your answer correct to two decimal places.
7 Bottles of wine contain 0.74 litres each. How much do 38 bottles contain? Give your answer correct to one decimal place.
8 A lottery win of £1000 is shared equally among eleven people. How much is each share? Give your answer correct to two decimal places.
9 Over a week, 4 cm of rain fell. What was the average rainfall per day? Give your answer correct to two decimal places.
10 A vat contains 37 litres of water. It is poured into 43 bottles so that each bottle contains the same amount of water. How much does each bottle contain? Give your answer correct to three decimal places.

Remember to save the rounding until the end!

Significant figures

We can round numbers to one decimal place, or to two decimal places, as well as to the nearest whole number. Can you now see how to sort out the problem of the running track at the beginning of the chapter?

If the distance is measured in metres, we are dealing with hundreds of metres, so we can round $\frac{1}{3}$ of it to the nearest metre. This gives 133 m, which is accurate enough for the purposes. (Rounding to the nearest 10 m is too inaccurate in this case.)

If the distance is measured in kilometres, we are dealing with a total distance that is a fraction of a kilometre, so it is not sensible to round to the nearest kilometre. It is better to round $\frac{1}{3}$ of the distance correct to three decimal places. This gives 0.133 km.

> 133.33. . .m rounded to the nearest whole number = 133 m
> 0.133 33. . .km rounded to three decimal places = 0.133 km
>
> 133 m = 0.133 km

These roundings represent the same level of accuracy. For 0.133 km, the digit in the third decimal place represents thousandths of a kilometre. But a thousandth of a kilometre is one metre. So these are equal.

> Rounding a quantity expressed in kilometres to three decimal places
>
> is the same as
>
> rounding the same quantity expressed in metres, to the nearest whole number.

In both cases we keep the first three digits, 1, 3 and 3. These are the first three **significant** figures. They are significant because they show the level of accuracy.

The first non-zero digit of a number is its **first significant figure**. In the numbers below, the first significant figure is 7. In some of these there are 0s *before* the 7. Note that we don't count these 0s as significant figures.

> 73   7.32   0.7389   0.071 99   0.007 383

Exercise 7.12

Write down the first significant figure of the following numbers.

1 73

2 230 000

3 1.288

4 0.822

5 0.002 451

The digit after the first significant figure is the **second significant figure**. In the numbers below, the first significant figure is 8, and the second significant figure is 6.

> 86   8.62   0.861   0.086 73   0.008 67

After the first significant figure, 0 *does* count as a significant figure. In the numbers that follow, the first significant figure is 8, and the second significant figure is 0.

801   8.05   0.807   0.080 002   0.008 09

And so on. The digit after the second significant figure is the **third significant figure**, even if it is 0.

Exercise 7.13

1 What is the second significant figure in these numbers?
   a 316   **b** 9.493   **c** 0.78   **d** 0.000 132
2 What is the third significant figure in these numbers?
   a 395   **b** 1.0635   **c** 0.780 33   **d** 0.000 107
3 For the number 120 372 write down
   a the first significant figure
   b the second significant figure
   c the third significant figure.
4 For the number 0.086 045 write down
   a the first significant figure
   b the second significant figure
   c the third significant figure.

Rounding to a given number of significant figures

Earlier in this chapter we rounded numbers to a given number of decimal places. The same ideas apply to significant figures.

● *To round a number **to one significant figure**, look at the second significant figure. If it is 5 or more, round up. If it is 4 or less, round down.*

Example   Round these numbers correct to one significant figure.
   a 1329   **b** 0.0378   **c** 3076

   a The second significant figure is 3, so round down to 1000.
   b The second significant figure is 7, so round up to 0.04
   c The second significant figure is 0, so round down to 3000.

● *In the last example the second significant figure was 0, so we rounded down. This example shows that we must count 0 as a significant figure, after the first.*

Exercise 7.14

Round these numbers to one significant figure.

1 4177   **3** 0.3947   **5** 0.03054
2 670000   **4** 2056000

● *To round a number **to two significant figures**, look at the third significant figure. If it is 5 or more, round up. If it is 4 or less, round down.*

All the numbers below are rounded to two significant figures.

Round 1258 up to 1300   (the third significant figure is 5)
Round 7.8437 down to 7.8   (the third significant figure is 4)
Round 0.002868 up to 0.0029   (the third significant figure is 6)
Round 0.0465 up to 0.047   (the third significant figure is 5
                                        – by convention, round up)

A similar procedure holds for rounding a value to any other number of significant figures. The values below are all rounded to four significant figures.

Round 91384 down to 91380   (the fifth significant figure is 4)
Round 3.4858 up to 3.486   (the fifth significant figure is 8)

Leaving in 0s

When we are rounding to four significant figures, and the fourth digit is 0, then leave it in to show the level of accuracy. This is what we did when rounding to a given number of decimal places.

Round 0.12304 down to 0.1230   (the fifth significant digit is 4, so round down)
Round 0.034298 up to 0.03430   (the fifth significant digit is 8 so round the 9 up to 10, which means that '29' rounds up to '30')

Leave the 0 to show the level of accuracy

Exercise 7.15

1 Round the following to two significant figures.
    a 49328   **b** 9.8744   **c** 0.07083   **d** 0.0001398
2 Round the following to three significant figures.
    a 341492   **b** 3856   **c** 0.013527
3 Round the number 127328 to
    a one significant figure   **b** two significant figures   **c** three significant figures

4 Round the number 0.005 623 to
   a one significant figure   **b** two significant figures   **c** three significant figures
5 Round the following to three significant figures.
   a 0.230 261   **b** 0.1296   **c** 1.396

Rounding calculations

If you have to give the answer to a calculation correct to a certain number of significant figures, you must do the rounding *after* the calculation, not before or in the middle.

Example   There are 326 workers in a factory, each paid £286 per week. What is the total weekly wage bill for the factory? Give your answer correct to two significant figures.

Multiply £286 by 326, obtaining £93 236. The third significant figure is 2, so round down.
   Correct to two significant figures, the total wage bill is £93 000.

● *If we had rounded at the beginning as well as the end, we would have worked out £290 × 330, which gives £95 700. Correct to two significant figures, the wage bill is £96 000. This answer is different!*

Exercise 7.16

In these questions, give the answers correct to the given number of significant figures.

1 456 × 933, correct to one significant figure
2 0.192 × 0.388, correct to two significant figures
3 2334 × 0.277, correct to two significant figures
4 30 992 ÷ 97, correct to three significant figures
5 85 ÷ 57 630, correct to three significant figures

> Remember to leave the rounding until the end!

6 What is the total cost of 378 books at £6.50 each? Give your answer correct to two significant figures.
7 What is the total weight of 73 cars, each weighing 866 kg? Give your answer correct to three significant figures.
8 A length of 2 m is divided into 11 equal parts. How long is each part? Give your answer correct to two significant figures.
9 A lottery win of £500 is shared between seven people. How much does each person get? Give your answer correct to two significant figures.
10 Anna's car uses 43 litres of petrol when travelling 537 kilometres. What is her fuel consumption, in km per litre of petrol? Give your answer correct to two significant figures.

Decimal places and significant figures

What's the difference between rounding to three decimal places and to three significant figures?

Look at the running track example at the beginning of this chapter.

The distance could be measured in metres or in kilometres.

133.333333 m or 0.133333333 km

When rounding these measurements, it is sensible to round them to the same number of *significant figures*. Then we get the same value.

133 m or 0.133 km.

So when we round a measurement to a certain number of significant figures, it doesn't matter what units we are using. If we round to a certain number of decimal places, it does make a difference.

Exercise 7.17

Remember the metric system of measurements!

1 A weight is found to be 0.012 854 kg. Express this weight in grams. Round both the weights to two significant figures. Convince yourself the results are the same.
2 A volume is found to be 2370 ml. Express this volume in litres. Round both the volumes to two significant figures. Convince yourself the results are the same.
3 A distance is measured as 649 m. Convert this to kilometres and to centimetres. Round all three distances to two significant figures. Convince yourself the results are the same.

SUMMARY

- To round a number to three decimal places, say, look at the digit in the fourth decimal place. If it is 5 or more, round up. If it is 4 or less, round down. At the half-way point of 5, always round up.
- If you round a number to three decimal places, say, and the digit in the third decimal place is 0, you must leave the 0 in the answer to show the accuracy.
- When some fractions are converted to decimals, the digits continue for ever. You can show this by a dot on top of the repeating digit, or by dots on top of a repeating pattern of digits. You may want to give the value correct to a certain number of decimal places.
- When doing calculations, you may have to give the answer correct to a certain number of decimal places. Save the rounding to the end of the calculation. Don't do it at the beginning or half-way through.
- The first significant figure of a number is its first non-zero digit. A digit after the first significant figure is the second significant figure, even if it is zero. The first significant figure can't be zero, but any of the next significant figures can be zero.
- You can round measurements to a certain number of significant figures. The result will not vary if different units are used.

Exercise 7A

1 What is the digit in the third decimal place of 1.234 57?
2 Round 3467 to the nearest 10.
3 Round 3.422 to two decimal places.
4 Round 3.215 to two decimal places.
5 Write $\frac{7}{9}$ as a decimal, using the dot notation. Round this value to three decimal places.
6 Evaluate 3.44×1.255, giving your answer correct to two decimal places.
7 In the number 2.304 87, what is the third significant figure?
8 Round 34.933 to two significant figures.
9 Evaluate $44 \div 963$, giving your answer correct to three significant figures.
10 The width of a washer is 0.936 cm. Convert this to millimetres. Round both values to two significant figures.

Exercise 7B

1 The weight of a ship is 2 348 723 kg. Write this correct to the nearest 1000 kg.
2 Round 0.0234 to two decimal places.
3 Round 4.340 42 to three decimal places.
4 Round 9.697 to two decimal places.
5 Write $\frac{5}{17}$ as a decimal, giving your answer correct to three decimal places.

6 A motorist pays £26 for 41 litres of petrol. Find the cost per litre, giving your answer correct to two decimal places.

7 In the number 0.002 054, what is the third significant figure?

8 Round 950 830 to three significant figures.

9 A farmer has 382 sheep, which weigh on average 57 kg. What is the total weight of the sheep? Give your answer correct to two significant figures.

10 A lorry weighs 2.748 tonnes. Write this weight in kilograms. Round both numbers to three significant figures. Do both answers mean the same?

Exercise 7C   Ma1

ANNA numbers are 4-digit numbers that read the same backwards as forwards, for example 3773. For every ANNA number there is a corresponding NAAN number, found by putting the inner two numbers on the outside. So for example, if 3773 is the ANNA number, then 7337 is the NAAN number.

1 Write down twenty ANNA numbers, and then write down their corresponding NAAN numbers.

2 For each pair of ANNA and NAAN numbers, find the difference. What do you notice?

Exercise 7D   Ma1

Some fractions, such as $\frac{1}{3}$ or $\frac{1}{9}$, repeat after only one digit. Other fractions, such as $\frac{1}{11}$, repeat after two digits. Some, like $\frac{1}{7}$, repeat after six digits. Is there a rule?

Express these fractions as decimals. Is there any rule for when they repeat? Try some more fractions.

$$\frac{2}{9} \quad \frac{3}{7} \quad \frac{2}{11} \quad \frac{1}{99} \quad \frac{1}{33} \quad \frac{1}{999} \quad \frac{1}{9999} \quad \frac{1}{111}$$

You could show your results on a class poster.

8 Enlargement and similarity

Enlargement

On holiday you took this photograph of your sister. She likes it so much that she wants a larger copy for herself. You take it to a shop for it to be made larger.

Now, your sister would be very annoyed if the photo were just stretched horizontally. It would look like this picture!

She would also be annoyed if it were just stretched vertically. It would look like this!

> After an enlargement, a figure has the same shape but a different size.

A proper **enlargement** should increase both horizontal and vertical directions by the same amount. The final result should look like this. Your sister is the same shape in the enlargement as she is in the original picture!

> When you enlarge a shape, you must make sure that the same number is used to multiply all the lengths. Otherwise the enlarged shape will be distorted.

If you have a picture or shape that you want to make bigger, you **enlarge** it. Each length is multiplied by the same amount.

● *In an enlargement, the number which multiplies the lengths is the scale factor.*

In the diagram below, the sides of triangle B are twice as long as the sides of triangle A. So the scale factor is 2.

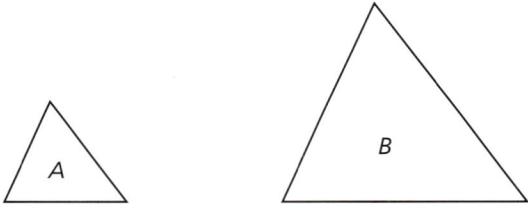

Example   In this diagram, a rectangle has been enlarged. Find the scale factor of the enlargement.

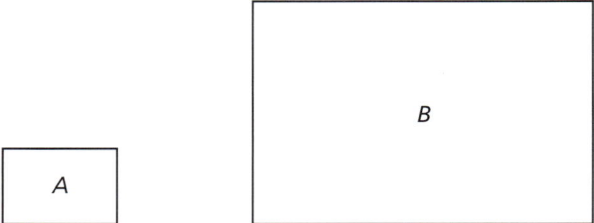

Measure the sides of both rectangles. Rectangle A is 1 cm by 1.5 cm, and rectangle B is 3 cm by 4.5 cm. The sides of rectangle B are three times the sides of rectangle A.

The scale factor is 3.

Exercise 8.1

1 This diagram shows two triangles. One is an enlargement of the other. What is the scale factor?

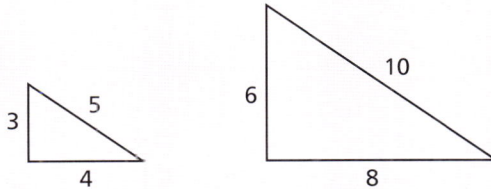

2 This diagram shows two rectangles. One is an enlargement of the other. Measure the lengths of the sides and find the scale factor of the enlargement.

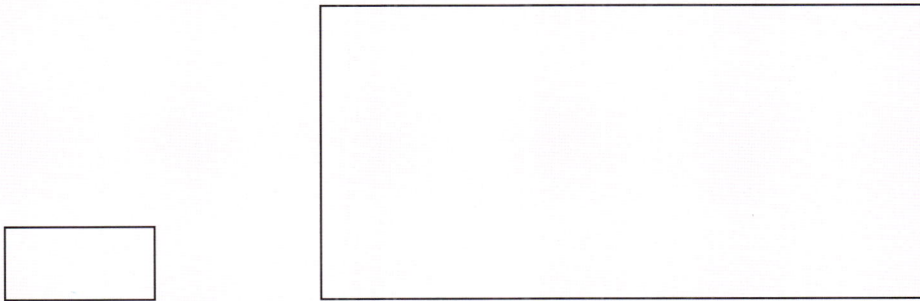

3 This diagram shows two triangles on a coordinate grid. One is an enlargement of the other. Find the scale factor of the enlargement.

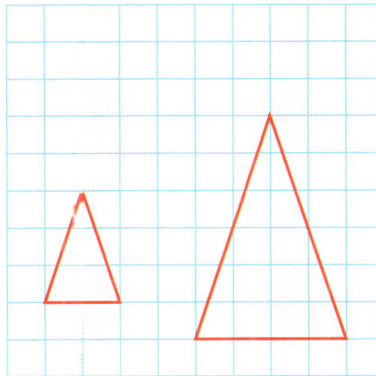

4 The two Xs below are printed in 36 pt and 72 pt. The right-hand X is an enlargement of the left-hand X. What is the scale factor?

> Typesetters measure the size of letters in *points* (pt). There are about 72 points in one inch.

X X

5 Set out a coordinate grid on graph paper or squared paper, with x- and y-axes going from 0 to 10. Plot two triangles: the first with vertices at (1, 1), (1, 3) and (2, 1), and the second with vertices at (4, 1), (4, 7) and (7, 1). One triangle is an enlargement of the other. Find the scale factor of the enlargement.

6 On the grid of question 5 (or a new grid) plot two rectangles. The first has vertices at (1, 1), (2, 1), (2, 4) and (1, 4), and the second has vertices at (3, 2), (5, 2), (5, 8) and (3, 8). Find the scale factor of the enlargement which takes the first rectangle to the second.

7 Look at the photos at the beginning of the chapter. What is the scale factor of the enlargement?

8 This diagram shows two pentagons. Measure the lengths of the sides. Explain why one pentagon is an enlargement of the other.

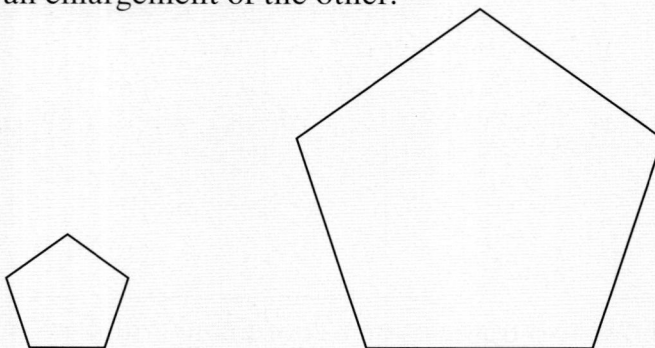

9 Triangle LMN is an enlargement of triangle ABC. Triangle ABC and the line LM are drawn below. Copy the diagram and draw N on it.

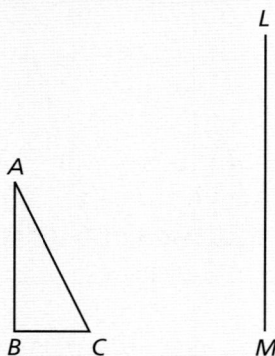

10 Rectangle ABCD is enlarged to rectangle PQRS. ABCD and PQ are drawn for you below. Copy the diagram and complete PQRS.

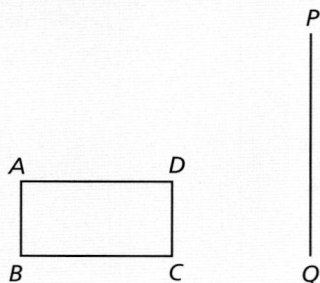

Exercise 8.2 Ma1

Paper comes in different sizes! Some common sizes are A3, A4, A5, and so on. Get hold of different sized sheets of paper, and measure their sides. Can you find two sheets for which one sheet is an enlargement of the other? What is the scale factor?

Doing enlargements

Simple shapes can be enlarged just by measuring the lengths of the sides and multiplying by the scale factor. If the shape is on a grid then the measurement is done for you. Just count out the lengths and multiply them by the scale factor! Then draw the enlargement.

Example Make an enlargement of the F shape shown below, with a scale factor of 2.

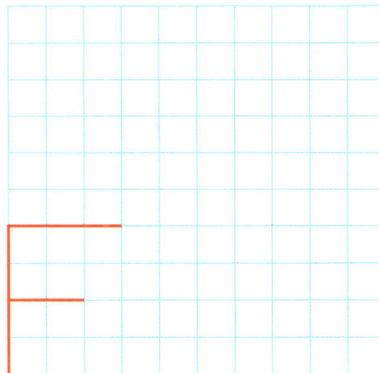

The top bar of the F is 3 units long, the middle bar is 2 units long, and the vertical line is 4 units long. Double all these lengths: draw an F shape with top bar 6 units, middle bar 4 units and with a vertical line of 8 units.

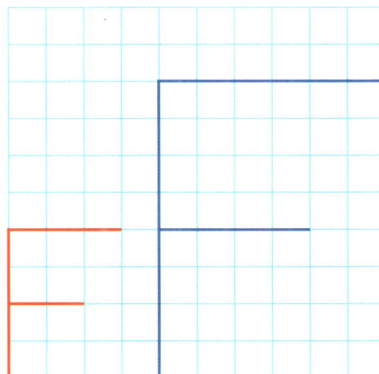

Exercise 8.3

For questions 1 to 3 copy the diagrams on to squared paper or graph paper, and enlarge them by a scale factor of 2. Draw the enlargement next to the original shape.

1

2

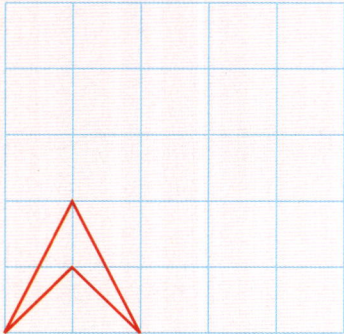

Make sure you leave enough space on the squared paper to fit the enlargement in!

3

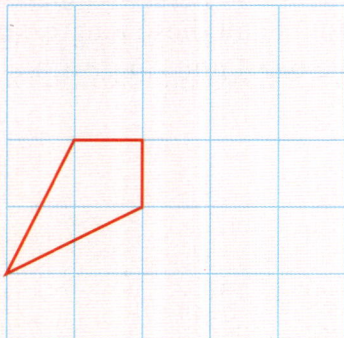

4–6 For questions 4 to 6, repeat questions 1 to 3, but enlarge the diagrams by a scale factor of 3.

7 Make an enlargement of the E shape below, with a scale factor of 2.

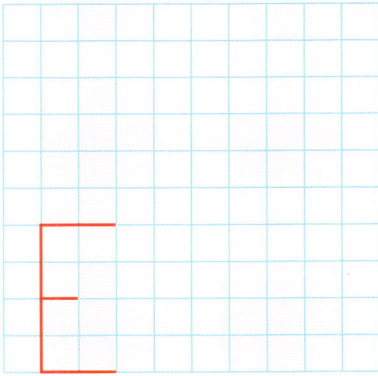

8 Below is a child's drawing of her mother. Enlarge it by a scale factor of 3.

9 Draw a coordinate grid, with *x*- and *y*-axes going from 0 to 10. Plot a triangle with vertices at (1, 1), (3, 1) and (1, 4). Make an enlargement of the triangle with a scale factor of 2.

10 On the grid of question 9 (or on a fresh grid) draw a rectangle with vertices at (0, 1), (0, 2), (2, 2) and (2, 1). Make an enlargement of the rectangle with scale factor 3.

Centre of enlargement

Daniel has cut out a shape and is holding it in front of a wall. Rachel is shining a torch onto it, so that the shadow falls on the wall. You can see that the shadow is an enlargement of the original shape. What is the scale factor?

You can try it with a real torch and a real wall, but here we shall just use a diagram. The torch is at T, and the shape is just a straight line AB. Draw straight lines from T, so that the shadow of AB falls at $A'B'$ on the wall.

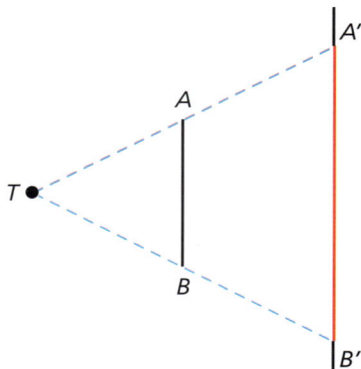

Here the shape on the wall is twice as far from the torch as the original shape. We can see that $A'B'$ is twice as long as AB.

Exercise 8.4

Copy the diagram below, making sure that distances measured are accurate. Following the example above, draw the shadow of the line AB on the wall when a torch is shone from point T.

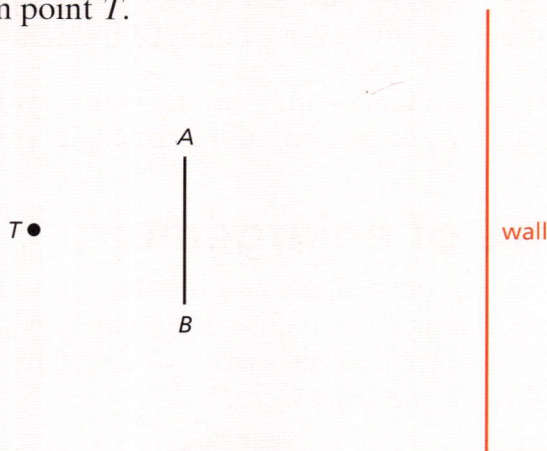

1 What are the distances of A and A' from the torch?
2 What is the scale factor of the enlargement?

Remember ratios from chapter 1.

You should find that the ratio of the distances from the torch gives the scale factor of the enlargement.

$$\frac{TA'}{TA} = \text{scale factor}$$

Putting it another way, the distances from the torch have all been multiplied by the scale factor of the enlargement.

$$TA' = TA \times \text{scale factor}$$

All the distances are measured from the torch at T. This point is the **centre of enlargement**. The process can be done on paper, without pretending there's a torch! Multiply all distances from the centre of enlargement by the scale factor.

Example   The triangle ABC is to be enlarged by a scale factor of 2. The centre of enlargement is at X.

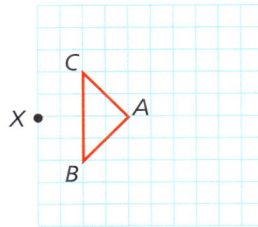

We need to measure the distance from the centre X to each point (**vertex**) on the shape. Multiply these distances by 2, then measure out the doubled distances from the centre. As ABC is on a grid, we can find and measure the distances just by counting out gridlines.

Point A of the triangle is 4 squares to the right of X. So the equivalent point, A', on the enlarged triangle will be $4 \times 2 = 8$ squares to the right of X. Plot the point A'.

The bottom point B of the triangle is 2 'diagonals' from X. The equivalent point, B', on the enlargement will be 4 diagonals from X. Plot the point B'.

The top point C of the triangle is 2 'diagonals' from X. The equivalent point, C', on the enlargement will be 4 diagonals from X. Plot the point C'.

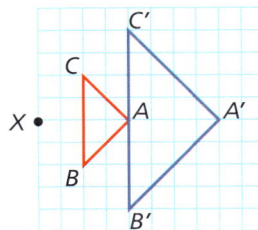

Remember:

A vertex (plural vertices) is a point where edges meet.

The new triangle $A'B'C'$ is an enlargement of ABC, with scale factor 2.

Exercise 8.5

For questions 1 to 3, copy the diagrams onto centimetre squared paper. Enlarge the shapes by a scale factor of 2, with centre of enlargement at *X*.

1

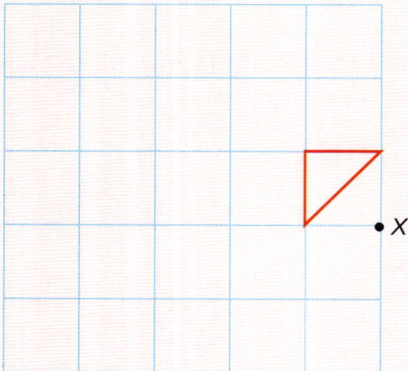

2

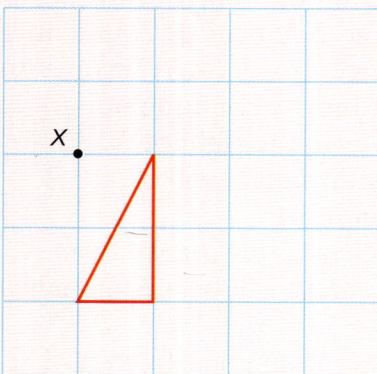

3

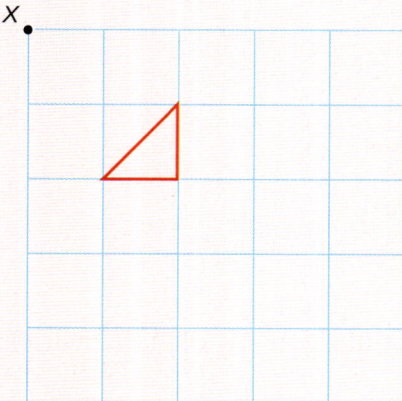

For questions 4 to 6, copy the diagrams onto $\frac{1}{2}$ cm squared paper. Enlarge the shapes by a scale factor of 3, with centre of enlargement at X.

4

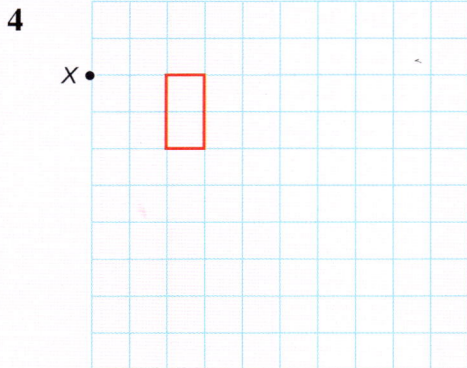

5

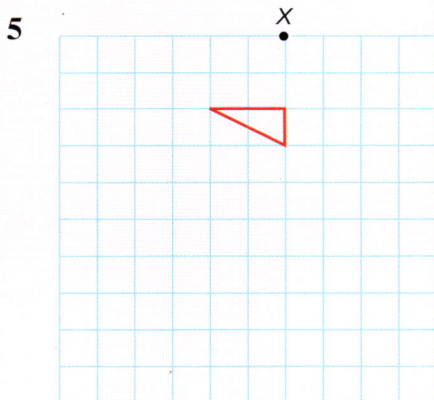

6

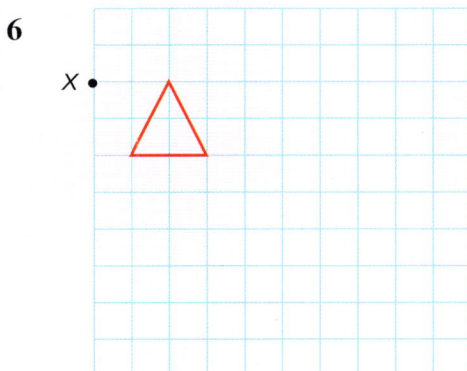

7 Draw a coordinate grid, with x- and y-axes going from 0 to 10. Plot a triangle with vertices at $(1, 1)$, $(1, 3)$ and $(4, 1)$. Make an enlargement of the triangle with a scale factor of 2, with centre of enlargement at the origin $(0, 0)$.

8 On the grid of question 7 (or on a fresh grid) draw a triangle with vertices at $(3, 3)$, $(1, 3)$ and $(3, 1)$. Make an enlargement of the rectangle with scale factor 3, with $(1, 1)$ as the centre of enlargement.

The centre of enlargement does not have to be outside the shape. It can be inside! Think of a lampshade – if there is a pattern on the lampshade, the light inside will cast a shadow of the pattern onto the walls of the room. The shadow is an enlargement of the pattern.

Example In the diagram below, the centre of enlargement X is inside the triangle ABC. Enlarge the triangle by a scale factor of 3.

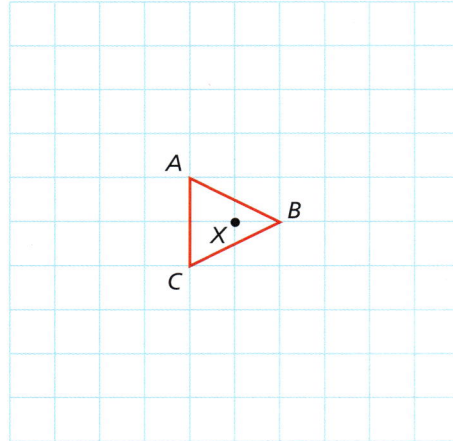

As before, measure the distance from X to each of the vertices of the triangle. As the triangle is drawn on a grid, we need only count gridlines. Multiply these distances by 3, and then measure out the new distances from X. For example:

> point C is 1 square down and 1 square to the left of X.
> Hence C' is 3 units down and 3 units to the left of X.

The final triangle $A'B'C'$ is shown below.

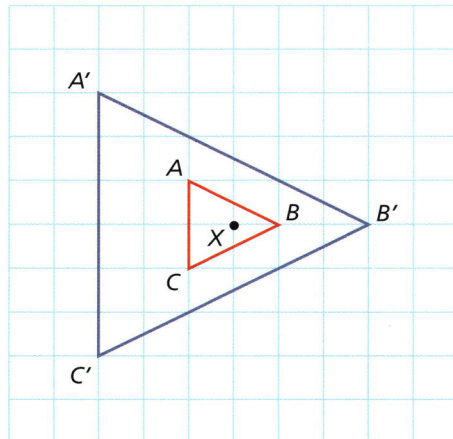

Exercise 8.6

For questions 1 to 3, copy the diagrams onto centimetre squared paper. Enlarge the shapes by a scale factor of 2, with centre of enlargement at X.

1

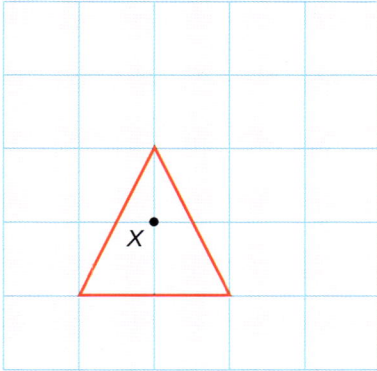

2

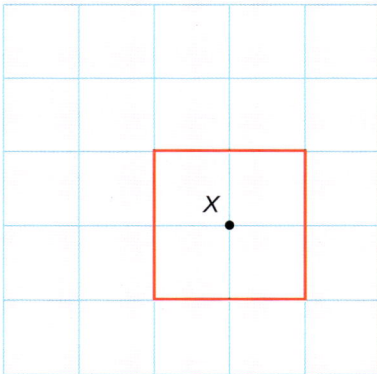

3

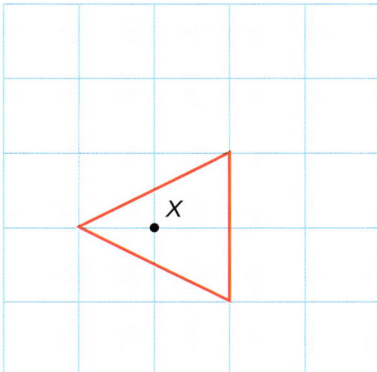

For questions 4 to 6, copy the diagrams onto squared paper with gridlines 0.5 cm apart. Enlarge the shapes by a scale factor of 3, with centre of enlargement at *X*.

4

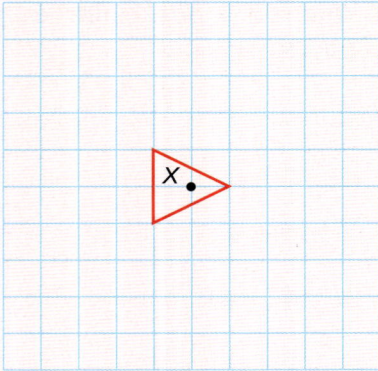

5

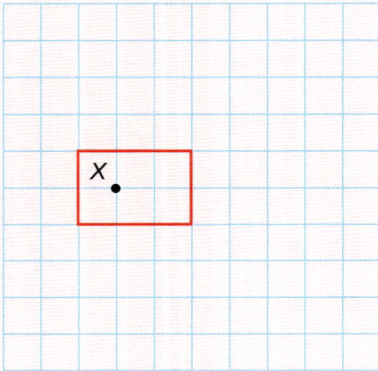

6

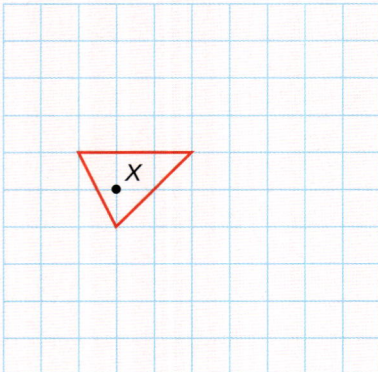

7 Set out a coordinate grid, with *x* and *y* going from 0 to 15. Plot a square with vertices at (8, 10), (10, 8), (8, 6) and (6, 8). Make an enlargement of the square with a scale factor of 3, and centre of enlargement at (8, 8).

8 On the grid of question 7 (or on a fresh grid) draw a triangle with vertices at (7, 8), (8, 6) and (7, 7). Make an enlargement of the triangle with scale factor 2, and (7, 7) as the centre of enlargement.

Maps

You have already met enlargements, in the scale diagrams of chapter 1. If you have a scale diagram of a rectangular field, the real field is an enlargement of the diagram. The ratio of one side of the diagram to the corresponding side of the field is the scale factor of the enlargement. A **map** is a familiar form of scale diagram. We use maps to find an unfamiliar street, or to plan a journey.

There are different scale maps available. The Ordnance Survey publishes a series of Landranger maps, covering the whole of Britain, in the scale 2 cm to 1 km. This is the same as 1 : 50 000.

Example   This map shows part of France. The scale is 1 cm to 2 km. Write this as a ratio.

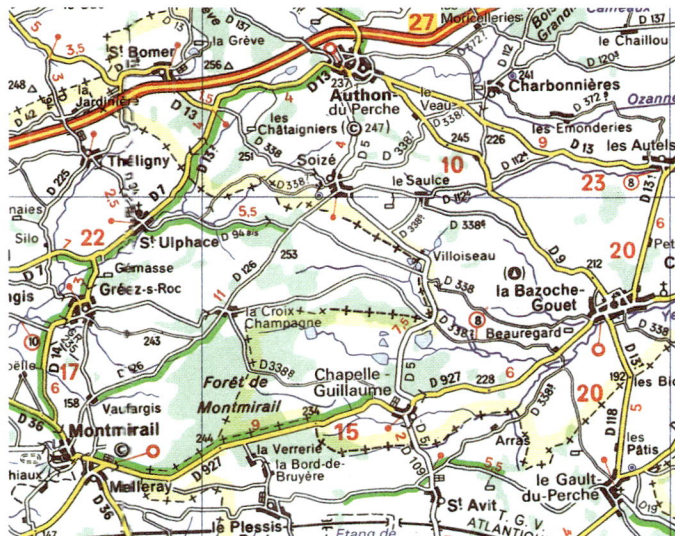

2 km is equal to 2000 m, and hence to 200 000 cm.
The scale is 1 : 200 000.

On the map above, you will find la Bazoche-Gouet near the right edge, and Authon-du-Perche near the top edge. What is the straight-line distance between these towns?

On the map, the distance is 4.5 cm. Each centimetre represents 2 km, so the real distance is 2 × 4.5 km.
The distance is 9 km.

Exercise 8.7

In questions 1 to 4, use the map on page 141 to work out the straight-line distance between the places given.

1 Authon-du-Perche and le Gault-du-Perche (near the bottom right corner)
2 la Bazoche-Gouet and Montmirail (near the bottom left corner)
3 Montmirail and St Bomer (near the top left corner)
4 le Gault-du-Perche and Montmirail
5 This map (right) is at the scale 1 cm to 10 km. Write this scale as a ratio.

For questions 6 to 8 use the map in question 5 to work out the distances between the places given.

6 Chartres and Orléans
7 Chartres and Blois
8 Orléans and Blois
9 The scale of a map is 4 cm to 1 km. Write this scale as a ratio.
10 The scale of a map is 1 : 1 000 000. Write this in terms of cm to km.

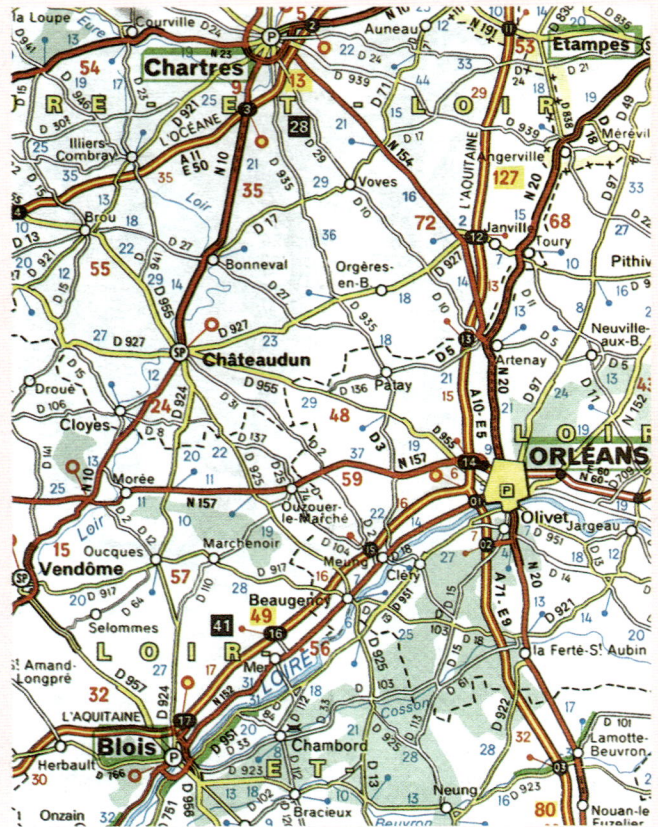

Similarity

After an enlargement, the figure changes size, but is still the same shape. We say that the two figures are **similar** to each other. A scale diagram of a building is similar to the real building. A map is similar to the land it covers.

● *If two figures are similar, then they have the same angles. The angles do not get multiplied by the scale factor!*

Because the earth is curved, any map of a country on a flat sheet of paper is only *approximately* similar to the real country. For a large country, there are many different ways of drawing a map. Here are some!

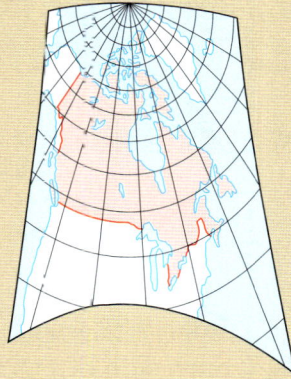

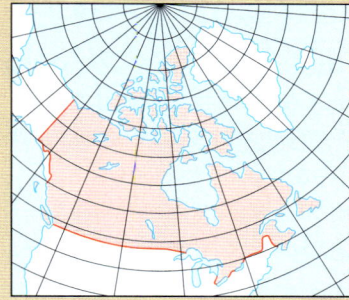

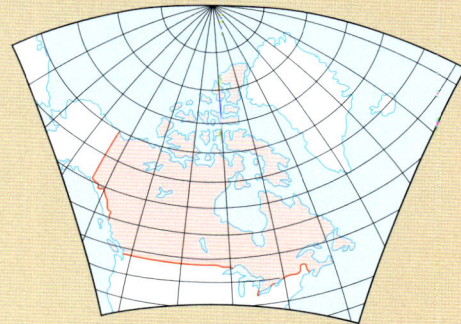

Exercise 8.8

1 This diagram shows a pair of similar triangles. Measure the angles of both triangles, and check that they are the same.

2 On squared paper or graph paper set out a coordinate grid, with x and y going from 0 to 10. Plot two triangles: the first has vertices at (1, 1), (1, 3) and (2, 1), and the second has vertices at (3, 3), (3, 9) and (6, 3). Check that the two triangles are similar. Measure the angles of the triangles, and check that they are the same.

3 If two triangles have the same angles, then they are similar. Is this true for quadrilaterals?

In questions 4 to 9, read the statement and decide whether it is true or false. If you think it is false draw a sketch to show that it is false.

4 All equilateral triangles are similar.
5 All rectangles are similar.
6 All regular pentagons are similar.
7 All rhombuses are similar.
8 All squares are similar.
9 All circles are similar.

If you've forgotten these figures, look back at chapter 4.

Suppose that we enlarge triangle *ABC* to triangle *DEF*, by a scale factor of 2. Then each side of *DEF* is twice the corresponding side of *ABC*. The ratios of the *DEF* sides to the *ABC* sides are all equal to 2.

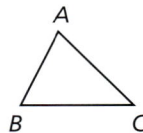

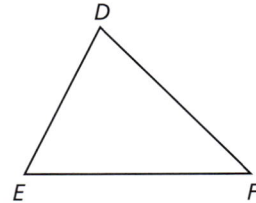

$$\frac{DE}{AB} = \frac{EF}{BC} = \frac{FD}{CA} = 2$$

If triangle *DEF* is similar to triangle *ABC* then, whatever the scale factor, the ratios between the corresponding sides are the same.

$$\frac{DE}{AB} = \frac{EF}{BC} = \frac{FD}{CA}$$

So, for example, if we know *DE*, *AB* and *BC*, then we can find *EF*. Note that the letters for each side of triangle *DEF* must come in the corresponding order to those of triangle *ABC*.

Example   This diagram shows triangles *XYZ* and *LMN*, which are similar. $XY = 5\,\text{cm}$, $YZ = 4\,\text{cm}$ and $MN = 20\,\text{cm}$. Find *LM*.

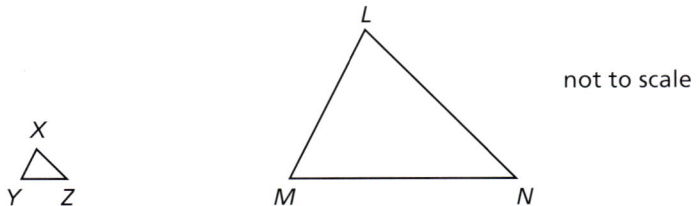

not to scale

The triangle *XYZ* has been enlarged to *LMN*. The side *XY* goes to *LM* and *YZ* goes to *MN*. We know *YZ* is 4 cm and *MN* is 20 cm. So the scale factor of the enlargement is 20 ÷ 4, that is, 5.
   Hence *LM* is 5 × 5 cm, that is, 25 cm.

Exercise 8.9

In questions 1 to 5 pairs of similar triangles are shown. In each case find the unknown side.

1

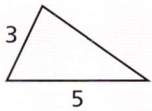 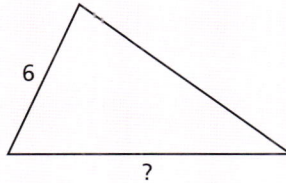

2

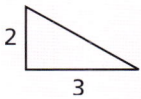 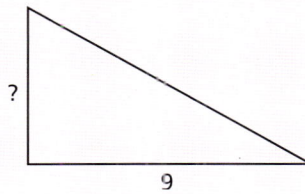

3

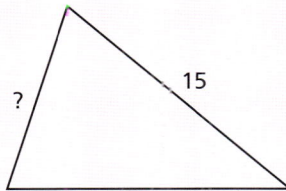

4

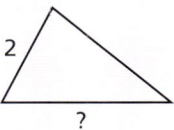 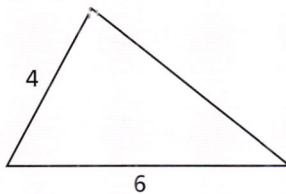

5

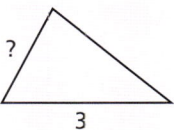 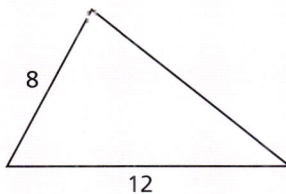

You can use similar triangles to solve practical problems.

How can you find a tree's height? It might not be safe to climb to the top. Wait for a sunny day, and measure the length of its shadow. Then get a stick about 2 m long. Plant it vertically in the ground so it is exactly 2 m high, then measure the length of its shadow.

Suppose the shadow of the tree is 20 m, and the shadow of the stick is 4 m. We can show this as on the diagram below. The triangles *ABC* and *DEF* are similar, so *ABC* is an enlargement of *DEF*. What is the scale factor?

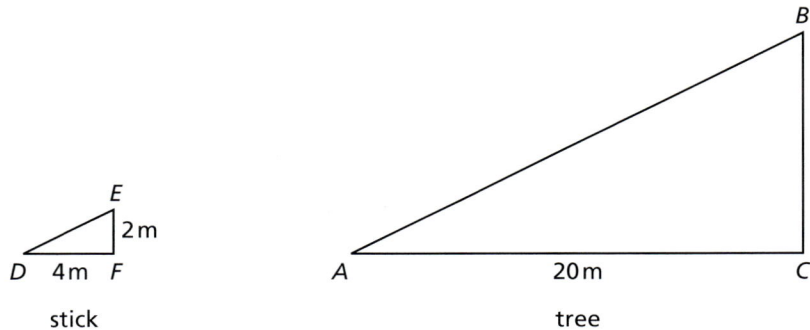

stick   tree

We have measured *CA* as 20 m, and *FD* as 4 m. Hence the scale factor of the enlargement is $20 \div 4 = 5$.

We know that the stick *EF* is 2 m. Multiply this by the scale factor to get the height of the tree, *BC*.

The tree is 10 m high.

Exercise 8.10

1 On a sunny day, the shadow of a tree was 36 m long. The shadow of a 2 m stick was 3 m long. What is the height of the tree?

2 This diagram shows a beam leaning against a wall, and resting on a support 1 m high. The foot of the beam is 2 m from the base of the support, and 6 m from the base of the wall. How high up the wall does the beam reach?

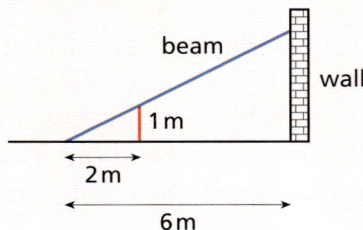

3 The diagram below shows a roof truss. A vertical support of height 0.8 m is 2 m from the gutter. The centre of the roof is 8 m from the gutter. How high is the roof ridge above the gutter?

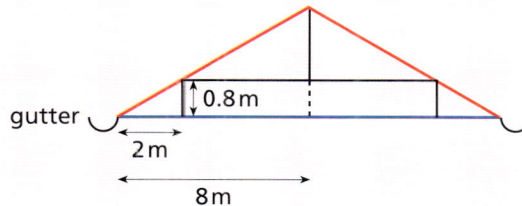

4 Jason wants to find the height of a tall tower. 40 m in front of the tower is a wall 3 m taller than his height of 1.5 m. He walks backwards for 10 m, and finds that the top of the tower and the top of the wall are in a straight line. What is the height of the tower?

SUMMARY

- After an **enlargement** a figure has the same shape, but a different size.
- An enlargement multiplies the lengths of a figure by a constant factor, called a **scale factor**.
- You can do enlargements by selecting a point as the **centre of enlargement**. Distances from the centre to the original figure are multiplied by the scale factor to produce the enlargement.
- With a map, the real object is an enlargement of the diagram.
- If one figure is enlarged to another, then the two figures are **similar**. Similar figures have the same shape but different sizes. The angles in similar figures are the same.
- If you have two similar triangles, you can solve problems about the lengths of their sides.

Exercise 8A

1 The leaf *A* is an enlargement of the leaf *B*. Find the scale factor of the enlargement.

B   *A*

2 Make a copy of the diagram below, on squared paper (1 cm gridlines). On it enlarge the T shape by a factor of 2.

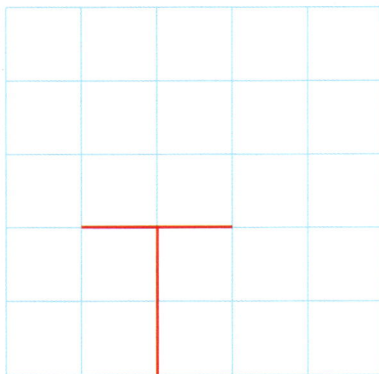

3 The diagram below shows two rectangles. Measure their sides. Explain why the right-hand rectangle is *not* an enlargement of the left-hand rectangle.

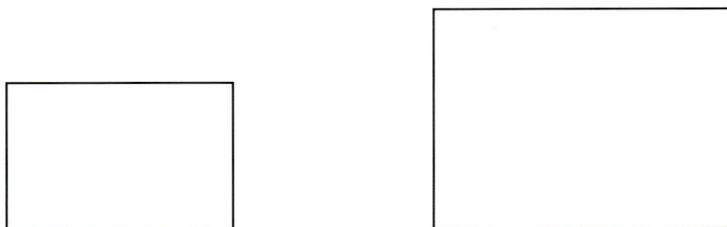

4 Copy the diagram below onto squared paper (0.5 cm gridlines). Enlarge the triangle *ABC* by a factor of 3, using *X* as the centre of enlargement.

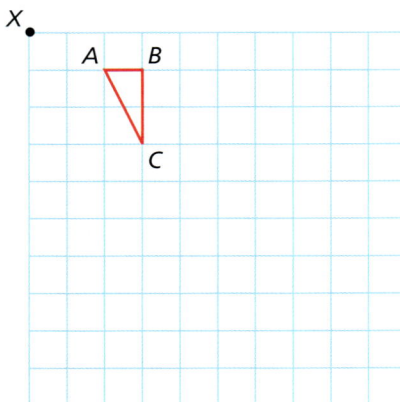

5 Set out a coordinate grid, with the axes going between −6 and 6. Plot a quadrilateral *ABCD* at (2, −2), (3, 1), (−3, 1) and (−2, −3). Enlarge *ABCD* by a scale factor of 2, using the origin (0, 0) as the centre of enlargement.

6 A map is in the scale 1 : 200 000. Two towns are 40 km apart. What is their distance apart on the map?

7 The map to the right shows part of Greece. The scale is 1 cm to 30 km. Write this scale as a ratio.

8 Use the map in question 7 to find the direct distance between Lárissa and Vólos. (Lárissa is near the top edge, and Vólos is near the right edge.)

9 Triangles ABC and LMN are similar. $AB = 8$ cm, $BC = 6$ cm and $MN = 9$ cm. Find LM.

10 The shadow of a tall building is 60 m long. At the same time, the shadow of a vertical 3 m stick is 9 m long. What is the height of the building?

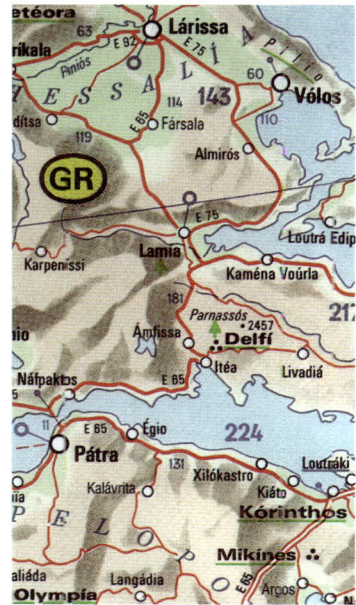

Exercise 8B

1 In the diagram below, the triangle ABC is an enlargement of LMN. Find the scale factor of the enlargement.

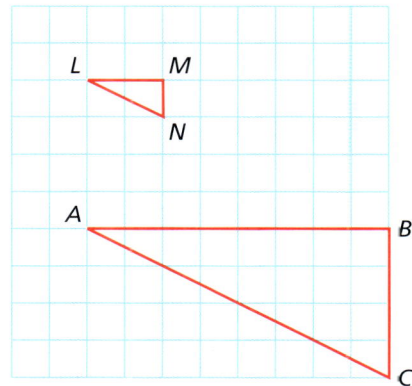

2 Copy the diagram below onto squared paper, and enlarge the triangle by a factor of 3.

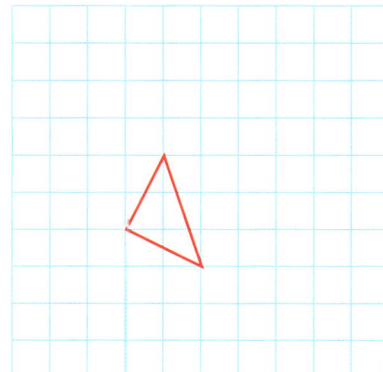

3 The diagram below shows two triangles which are similar to each other. Check that their angles are the same.

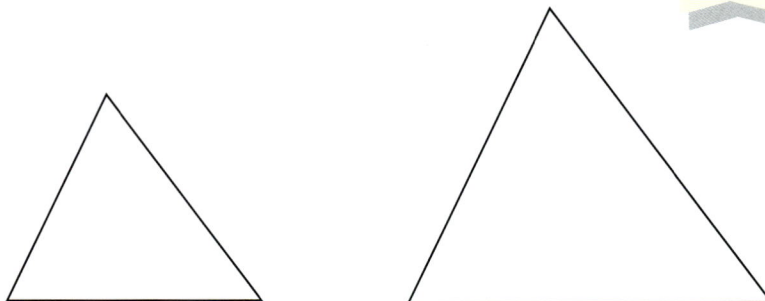

You will need:
• a protractor

4 Make a copy of the diagram below on squared paper. Enlarge the triangle ABC by a factor of 2, with X as the centre of enlargement.

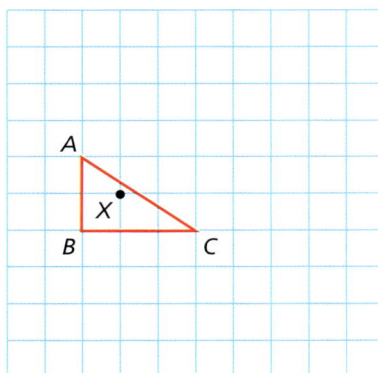

5 On holiday, Stefan takes a photo of a cathedral. In the photo, the cathedral is 2 cm high. He has the photo enlarged, so that the cathedral is now 3 cm high. What is the scale factor of the enlargement?

6 In the original photo of question 5, the width of the cathedral was 5 cm. What is the width in the enlarged photo?

7 Ordnance Survey Landranger maps are at the scale 2 cm to 1 km, or 1:50000. The scale is also given as $1\frac{1}{4}$ inches per mile. Find this scale as a ratio. How close is the scale to 1:50 000?

Remember:

There are 12 inches in a foot, 3 feet in a yard, and 1760 yards in a mile.

8 The map in question 7 of exercise 8A has a scale of 1 cm to 30 km. I start from Vólos and go South for 150 km. Where do I end up?

9 Rectangles $ABCD$ and $WXYZ$ are similar. $AB = 15$ cm, $AD = 20$ cm and $WX = 1\frac{1}{2}$ cm. Find WZ.

10 The diagram below shows a roof truss. The width from *A* to *B* is 5 m. The distance from *A* to *C* is 2 m, and the height of the post *CE* is 1 m. What is the height of the post *BD*?

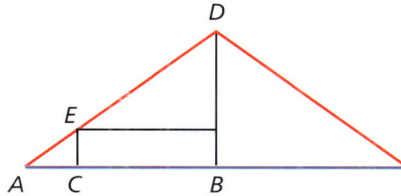

Exercise 8C   Ma1

Before computer technology, enlargements could be made using an instrument called a **pantograph**. You can make a simple pantograph using cardboard and paper-fasteners.

 Cut out four strips of stiff cardboard, two of length 20 cm and two of length 10 cm. Make holes at both ends of all the strips, and in the middles of the longer strips. Use paper-fasteners to join the strips together as shown. Make sure that the strips can rotate freely about each other.

You will need:

• cardboard
• scissors
• ruler
• paper-fasteners
• drawing pin

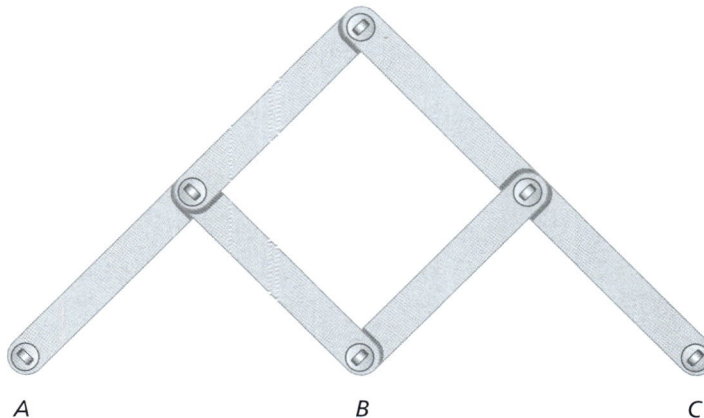

 Suppose you want to enlarge a shape. Fix the end *A* a short distance from the shape with a drawing pin. Move point *B*, where the shorter bits meet, to trace round the shape. While this is happening, hold a pencil at *C* to mark out another shape.

1 What have you done?
2 What happens if you move *C* to trace round the shape, and put the pencil at *B*?
3 What if you fix *B*, move *A* round the shape and put the pencil at *C*?

Exercise 8D 〔Ma1〕

With modern technology there is no need to do enlargements by hand. You can use an ordinary office photocopier to enlarge shapes.

You will need:
• photocopier

   You will probably find that the enlargement is measured in percentages. An enlargement of 150%, for example, corresponds to a scale factor of 1.5. Enlarge a picture by 150%, and check that all lengths have been multiplied by 1.5.

1 What happens if you enlarge *twice* by a scale factor of 1.5?

2 What is the scale factor?

An enlargement of a scale factor of 2 is an enlargement of 200%. Probably the enlargement will not go up as high as this. That doesn't matter. You can achieve an enlargement with a scale factor of 2 by a combination of two other enlargements.

3 What should they be?

9 Probability

Your sister took her driving test this morning. How do you think she did?

I don't know, but she must have either passed or failed, so I suppose the probability that she passed is $\frac{1}{2}$.

It's better than that. About 60% of 17 year olds pass first time, so that's her probability, 0.6.

But she's already taken the test, so she's either passed or failed. The probability is either 0 or 1, we just don't know which yet.

Who is right?

Probability of events

Some things are certainly going to happen, some things are certainly *not* going to happen. Some things may or may not happen. Consider the following future events.

> The sun will rise tomorrow.
> It will snow near your home next Christmas.
> England will win the next cricket test match against India.
> There will be two Tuesdays next week.
> When I spin a fair coin, it will come up Heads.
> When I throw a drawing pin on the floor, it will land with its point upwards.

To show how likely we think an event is, we give it a number. This number is the **probability** of the event happening. If we are **certain** that something will happen, we give it the number 1. If something is certain *not* to happen, that is, it is impossible, we give it the number 0. From the list above, the first event is certain, and the fourth event is impossible.

> The probability that the sun will rise tomorrow is 1.
> The probability that there will be two Tuesdays next week is 0.

These are shown on the probability line below. Certain events are on the right, at 1. Impossible events are on the left, at 0.

> The probability of a certain event is 1. The probability of an impossible event is 0.

impossible   certain

0   1

But what about those events which may or may not happen? How likely do you think the third event in the list above is? That will depend on your own opinion – a keen England cricket fan might think it likely that England will win. But a fan of the Indian team may think that India is more likely to win. The probability given to the event will depend on personal opinion.

> All probabilities can be expressed as a number (fraction) between 0 and 1.

Graham thinks that the probability that England will win is $\frac{2}{3}$. Niraj thinks that the probability that England will win is $\frac{1}{4}$.

You can see that they put the event at different places on the probability line below.

Niraj   Graham

$\frac{1}{4}$ $\frac{2}{3}$

0   1

Exercise 9.1

Questions 1 to 7 each state an event. Guess the probabilities of these events. Make a copy of the probability line below, and mark the position of your guesses on the line. Ask a friend to do the same, and see whether or not you agree.

0   1

1 An Englishwoman will win the Wimbledon tennis tournament next year.
2 The Labour Party will win the next general election.
3 It will rain tomorrow.
4 Next August will contain 30 days.
5 When you throw a dart at a dartboard, it will land in the bullseye.
6 You will pass your driving test the first time you take it.
7 Next week there will be seven days.

People state probabilities in other ways. Questions 8 to 10 give some of them.

8 Jacques says that he has a 30% chance of winning the tennis tournament. Express this percentage as a fraction.
9 Angela says she has a fifty-fifty chance of qualifying for the high jump at the county sports. What does she think her probability of qualifying is?
10 Rachel reckons she is twice as likely to finish her project as not to finish it. What does she think her probability of finishing is?

Theoretical probability

For many events, people don't always agree about their probabilities. There are some events, however, for which we can work out the probability by theory.

Example

A coin has Heads on one face and Tails on the other. If the coin is fair then, when it is spun, it is just as likely to come down Heads as it is to come down Tails. So Heads and Tails have equal probabilities of $\frac{1}{2}$.

There are other familiar cases.

Examples

Suppose a fair dice has the numbers 1 to 6 written on its faces. What is the probability that the number 5 will be uppermost when it is rolled?

When it is rolled. each of the faces is equally likely to come uppermost. So the probability that the number 5 will be uppermost is $\frac{1}{6}$.

If a standard pack of cards is well shuffled and the top card dealt out, what is the probability that it is the Queen of Spades?

There are 52 cards in a standard pack. One of these is the Queen of Spades, so the probability is $\frac{1}{52}$.

● *The general rule is this: if an experiment has n equally likely results, then the probability of each particular result is 1/n.*

The results must be *equally likely*. The coin must be fair, the dice must be fair, the pack must be complete and well shuffled. If you have a misshapen coin then you can't deduce the probability that it will come up Heads!

Dice aren't always fair! This photograph shows ancient Roman dice, which were often very misshapen!

Exercise 9.2

1 A fair tetrahedral (four-sided) dice has the numbers 1 to 4 on its faces. It is thrown up in the air. What is the probability that it lands on the face labelled 3?

2 A bag contains ten counters, labelled 1 to 10. One counter is taken out of the bag. What is the probability that it is labelled 7?

3 In the National Lottery, there are 49 balls labelled 1 to 49. What is the probability that the first ball drawn is number 23?

4 The diagram below shows an eight-sided spinner, with the letters A to H labelling its edges. When it is spun, what is the probability that it lands on the edge labelled E?

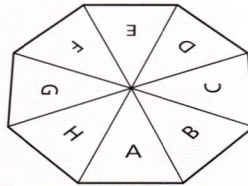

5 A roulette wheel has 36 holes labelled 1 to 36, and a hole labelled zero. What is the probability that the ball lands in the zero hole?

6 Every student at Stowell College has a 6-digit number. What is the probability that Robert's number ends with 9?

7 A day of the week is picked at random. What is the probability that Monday is picked?

8 Neela has a cash card. When she uses it to withdraw money, she has to type in a 4-digit PIN number. She remembers the first three digits, but has forgotten the last digit. If she picks a number at random, what is the probability that she gets it right?

9 A month is picked at random. What is the probability that August is picked?

10 Look again at the conversation at the beginning of this chapter. One of the girls argued that, as there are two possible results, each of them has probability $\frac{1}{2}$. Do you agree with her? Give a reason for your answer.

Favourable outcome

The standard pack of cards has 52 cards, which are divided into four suits: Spades, Hearts, Diamonds and Clubs. Each suit has 13 members: Ace, King, Queen, Jack, 10, 9, 8, 7, 6, 5, 4, 3 and 2.

Example If a standard pack of cards is shuffled well and the top card is dealt out, what is the probability that it is a Heart?

There are 13 Hearts out of a total of 52 cards. The probability is $\frac{13}{52}$.

Simplify this fraction by dividing tcp and bottom by 13.

$$\text{Probability of a Heart} = \frac{13}{52} = \frac{1}{4}.$$

Another way of solving the problem is this: there are four suits, each of which is equally likely to be dealt. So the probability that a Heart is dealt is $\frac{1}{4}$.

We wanted to know the probability of dealing a Heart. For us, this was a **favourable outcome**. Suppose we have several equally likely results. Count the number of results which give the favourable outcome, and divide by the total number of results. This gives the probability of the favourable outcome.

$$\text{probability} = \frac{\text{number of results which give favourable outcome}}{\text{total number of results}}$$

Example A bag contains 20 counters, of which 7 are red and 13 are green. If a counter is drawn out of the bag, what is the probability that it is red?

Here the favourable outcome is drawing a red counter. There are 7 of these, out of a total of 20.
The probability is $\frac{7}{20}$.

In general, if there are n equally likely results, and r of them give the favourable outcome, then the probability of the favourable outcome is $\frac{r}{n}$.

Example Suppose that a 'lucky dip' at a fair holds 100 packets, of which 67 contain prizes. You reach into the 'lucky dip' and pull out a packet. What is the probability that you win a prize?

Clearly it is a favourable outcome if the packet contains a prize. This happens in 67 out of the 100 possible results. So the probability of the favourable outcome is $\frac{67}{100}$.

Exercise 9.3

1 A card is drawn from a well-shuffled pack. What is the probability that it is a King?
2 A fair dice is rolled. What is the probability that an odd number is uppermost?
3 In a cupboard there are 12 similar tins, of which 5 contain soup. You reach in to the cupboard and pull out a tin. What is the probability that it is a soup tin?
4 The diagram below shows a nine-sided spinner, with the numbers 1 to 9 labelling its edges. When it is spun, what is the probability that it lands on an edge with an even number?

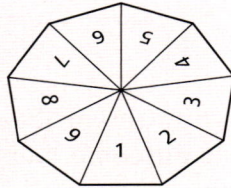

5 In the National Lottery, there are 49 balls, marked with the numbers 1 to 49. What is the probability that the first ball drawn will carry a single-digit number?
6 In the National Lottery, what is the probability that the first ball drawn will carry a number divisible by 7?
7 In the National Lottery, what is the probability that the first ball drawn will carry a number containing the digit 3?
8 A charity sells 2000 raffle tickets, for which there are 15 prizes. If I buy a ticket, what is the probability that I win a prize?
9 A month is picked at random. What is the probability that it contains fewer than 31 days?
10 A card is dealt from a well-shuffled pack. What is the probability that it is either a Heart or an Ace (or both)?

Experimental probability

A fair coin is one in which both sides are equally likely to come up. But what about a coin which is *unfair*, perhaps because it is misshapen? How can you find the probability that it will show Heads?

It would be impossible to find the theoretical probability. Instead, you must find it by experiment. Perhaps you spin the coin 100 times, and it shows Heads 40 times. Then your **experimental probability** for Heads is $\frac{40}{100}$, which simplifies to $\frac{2}{5}$.

This is only an experiment. Someone else might get a different result – perhaps they got a total of 35 Heads out of 80 spins, giving a probability of $\frac{35}{80}$, which simplifies to $\frac{7}{16}$.

Exercise 9.4 Ma1

Take a drawing pin, and drop it onto the floor many times. How often does it land with the point upwards? What is the probability that it lands with the point upwards? Copy and complete the tally chart below.

number of drops	number of landings with point upwards

Try with different shapes of drawing pin, some with long spikes and some with short spikes. Is there a connection between the length of spike and the probability that it lands with the point upwards?

Exercise 9.5

1 A supermarket display of 150 cans of beans toppled over, and 50 cans landed on a flat face. What is the experimental probability that a can will land on a flat face?

2 In a certain town, over the past 40 years, it has snowed on Christmas Day on 15 years. What is the experimental probability that it will snow on next Christmas Day?

3 From a group of 200 people, 34 are left-handed. What is the experimental probability that someone picked at random is left-handed?

4 A dice is known to be unfair. When it is rolled 80 times, it comes up six on 20 occasions. What is the experimental probability that this dice comes up six?

5 A misshapen coin is spun 100 times, and gives Tails 65 times. What is the experimental probability that this coin gives Tails?

6 A piece of buttered toast is dropped from the table 50 times, and lands butter side down 38 times. What is the experimental probability that the toast will land on its buttered side?

7 Barchester City and Coaltown United have played 48 matches against each other. Coaltown has won 27 of these matches. What is the experimental probability that Coaltown will win the next match?

8 Brian threw a dart at a dartboard 60 times, and it hit the bullseye 18 times. What is the experimental probability that Brian can hit the bullseye with a single dart?

9 A striker takes 30 penalty shots at goal, and the goalkeeper saves 8 of them. What is the experimental probability that the goalkeeper can save a penalty?

10 Darren bought a very cheap motorcycle. He rode it for 40 journeys and it broke down on 16 of them. What is the experimental probability that the motorcycle will break down on the next journey?

Predicting results

Suppose you spin a fair coin 100 times. You would expect the number of Heads to be $\frac{1}{2} \times 100$, that is, 50. You may not get this value – you may get 49 Heads or 51 Heads. You might get only 20 Heads.

If you repeat an event many times, you can predict the number of times you get a particular result. Multiply the probability of the result by the number of times you repeated the event. This is only a prediction – you may get more or fewer successful results, just by chance.

Example

You may be lucky, and win many more than three times. You may be unlucky, and not win at all. It is only a prediction!

In the National Lottery, the probability that an entry will win a prize is about $\frac{1}{70}$. If you have one entry per week for 210 weeks, how many prizes do you expect to get?

Multiply $\frac{1}{70}$ by 210, obtaining 3. You expect to win three times.

Exercise 9.6

1 A fair dice is rolled 90 times. How many sixes do you expect?
2 A fair coin is spun 120 times. How many Heads do you expect?
3 The probability that a 17-year-old will pass the driving test is 0.6. Out of a group of thirty 17-year-olds taking the test, how many do you expect will pass?
4 Over many years, it has been found that it will snow on 3 out of 10 Christmas Days. Over the next 20 years, how often do you expect it will snow on Christmas Day?
5 Namik finds that his probability of hitting the bullseye on a dartboard is $\frac{1}{8}$. If he throws 40 darts, how many bullseyes do you expect he will get?
6 The probability that a television will break down in a given week is 0.02. How often do you expect it will break down in 150 weeks?
7 A pizza shop finds that a third of the pizzas it sells are De Luxe. If it sells 180 pizzas, how many do you expect will be De Luxe?
8 A firm finds that a quarter of the letters it sends are delivered late. If it sends out 280 letters, how many do you expect will be delivered late?
9 A used car saleswoman finds that about one in every ten enquiries results in a sale. In a week she receives 80 enquiries. How many sales does she expect to make?
10 For the past 150 years the universities of Camford and Oxbridge have had an annual tiddlywinks match. Camford has won 85 of these matches. What is the experimental probability that Camford will win the next match? How many of the next 30 matches do you expect will be won by Camford?

Sum of probabilities

When you roll a fair dice, the probability that you will get a 6 is $\frac{1}{6}$. What is the probability that you *won't* get a 6? There are five other possible numbers, 1, 2, 3, 4 and 5. So the probability is $\frac{5}{6}$.

Do you notice anything about these fractions?

$$\frac{1}{6} + \frac{5}{6} = 1$$

They add up to 1!

So, to find the probability that you *won't* get a 6, just subtract from 1 the probability that you *do* get a 6.

This is true in other cases. If the probability that something will happen is p, then the probability that it *won't* happen is $1 - p$.

> Probability of *A* not happening
> = 1 − probability of *A* happening.

Suppose the probability that it will rain tomorrow is $\frac{1}{3}$. Then the probability that it *won't* rain tomorrow is $1 - \frac{1}{3} = \frac{2}{3}$.

Example   Jackie's sister is gloomy about her chances in the driving test.

My chance of passing is only $\frac{1}{4}$. So what's my probability of failing?

Subtract $\frac{1}{4}$ from 1, obtaining $\frac{3}{4}$.
Her probability of failing is $\frac{3}{4}$.

Exercise 9.7

1 The probability that it will rain next Midsummer's day is $\frac{1}{8}$. What is the probability that it won't rain?

2 When you throw a dart at a dartboard, the probability that you will score a bullseye is $\frac{1}{7}$. What is the probability that you won't score a bullseye?

3 In a raffle, the probability that a ticket will win a prize is $\frac{1}{20}$. What is the probability that it won't win a prize?

4 A fair spinner has eight edges, labelled 1 up to 8. What is the probability that it won't land on the edge labelled 8?

5 A misshapen coin is such that the probability of it landing Heads is $\frac{2}{3}$. What is the probability of it landing Tails?

6 The chance that Jon will be late for school tomorrow is $\frac{1}{9}$. What is the probability that he won't be late?

7 Claire plants an avocado stone. She is told that there is a $\frac{7}{10}$ chance that it will germinate. What is the probability that it won't germinate?

8 A by-election is held. The probability that the candidate of the Progressive Party will win is 0.6. What is the probability that the candidate doesn't win?

9 When Jo plays Stephen at tennis, the probability that Jo wins is 0.3. What is the probability that Jo doesn't win?

10 A misshapen dice is rolled 100 times, and comes up 6 for 36 rolls. What is the experimental probability that this dice gives a 6? What is the experimental probability that it doesn't give a 6?

SUMMARY

■ The **probability** of an event shows how likely we think it is.
■ A certain event has probability 1, and an impossible event has probability 0.
■ Different people may give different values for the probability of an event.
■ If there are n equally likely outcomes, the theoretical probability of each is $\frac{1}{n}$.
■ If you can't find a probability by theory, you have to do an experiment. The value found by the experiment may vary. It is only an estimate.
■ If you know the probability of an event, you can predict the number of times it will occur. But your prediction may not happen.
■ If the probability that an event will happen is p, then the probability that it won't happen is $1 - p$.

Exercise 9A

1 In a few years' time you will take Maths GCSE. What do you think is your probability of getting a grade A?

2 A fair octahedral (8-sided) dice has the numbers 1 to 8 on its faces. When it is rolled, what is the probability that it lands on the face labelled 5?

3 When playing the word game *Scrabble*, you pick a tile from a bag containing five tiles, with the letters A, C, J, X and Z on them. What is the probability that you pick X or Z?

4 You throw a knife in the air 40 times, and it lands on its point 15 times. What is the experimental probability that this knife will land on its point?

5 The octahedral dice of question 2 is rolled 200 times. How many times do you expect it will land on the face labelled 7?

6 A bag contains 32 blue balls, 16 red balls and 12 white balls. One ball is picked at random. What is the probability that it is not white?

7 A bag contains four red balls and six black balls. A ball is picked at random, and then put back in the bag. If this is done 100 times, how often do you expect to pick a red ball?

8 A pack of cards is incomplete. There are only 7 hearts, in 40 cards. If a card is drawn, what is the probability that it isn't a heart?

9 A fair icosahedral (20-sided) dice has the numbers 1 to 20 on its faces. When it is rolled, what is the probability that it doesn't land on the face labelled 17?

10 A misshapen coin is such that the probability of getting Heads is p. What is the probability of getting Tails?

Exercise 9B

1 Make a copy of the number line below, and on it place the probability of each of the following.

a March will follow February next year.
b Next year, October will come immediately after August.

2 On your copy of the number line in question 1, place the probability that you will be married before you are 21.

3 You are one of a group of five people who draw straws to see who will do the washing up. What is the probability that you will draw the short straw?

4 A box of chocolates contains seven with hard centres and eight with soft centres. If you pick one at random, what is the probability that it has a soft centre?

5 An opinion poll of 100 people found that 60 of them supported the return of capital punishment. What is the experimental probability that a person will support capital punishment?

6 In a constituency, 40% of the voters support the Labour Party. If you pick 200 people at random, how many Labour supporters do you expect to find?

7 There are 25 prime numbers less than 100. If I pick a number from 1 to 100 at random, what is the probability that it is not prime?

8 At the end of a football match, the home team has won, lost or drawn the match. If the probability that it wins is $\frac{1}{3}$, what is the probability that it loses or draws?

9 Martha and Anthea play 60 points at squash, and Martha wins 42 of them. What is the experimental probability that Anthea will win the next point?

10 There are four possible results for a completed cricket match: win, lose, draw or tie. The first three outcomes have probabilities $\frac{1}{4}$, $\frac{1}{3}$ and $\frac{2}{5}$, respectively. What is the probability of a tie?

Exercise 9C Ma 1

You will need:
• 5p coin
• paper
• pencil
• ruler

Take a 5p coin, and mark out a grid of squares which are large enough for the coin to fit entirely within a square. If you throw the coin onto the grid, it either lands entirely within a square or it crosses some of the lines. The diagram shows the various possible ways it could land.

Find the experimental probability that the coin lands entirely within a square. If someone else did the same experiment as you, would you both get the same result?

In fact, you can work out the *theoretical* probability that the coin will land entirely within a square. To find this you will have to measure the diameter of the coin and the side of the square. See if you can work out the theoretical probability. Is it close to your result for the experimental probability?

Exercise 9D Ma 1

If you have a scientific calculator, then it will probably have a button which generates 'random numbers'. It may be called 'RAN#'. Find this button, and press it a few times. Each time it should give a number, chosen at random, between 0 and 1.

What is the probability that the random number is greater than 0.7? Press the button several times, and count how often the number was greater than 0.7, on a copy of the tally sheet below.

number of presses	number of times value greater than 0.7

From your results find the experimental probability that the number is greater than 0.7. Get someone else to do the same experiment. Do you get the same result?

If the number is truly random, then you can find the theoretical probability of this event. Is your experimental result close to the theoretical result?

10 Organising data

Sean and Jacqui have been arguing about what makes a good swimmer. They question several people in their school, asking for their shoe size and the length they can swim without a break. They find that, on the whole, people with a larger shoe size can swim further.

That proves what I said. If you've got big flippery feet, that helps you to swim further

Not at all – lots of swimming makes your feet get larger!

Who is right?

You may have done an investigation which involved collecting lots of information in the form of numbers. A great list of numbers doesn't tell you very much. They have to be organised in some way before you can make sense of them.

Frequency tables

Suppose you have collected data for the ages of a group of forty people. The ages might be these.

```
23  27  33  40  55  52  58  43  48  30
38  29  25  36  41  44  37  50  27  52
20  43  39  36  47  48  42  35  31  56
58  42  27  33  30  41  53  39  22  25
```

You can't make sense of these numbers just by looking at them. No one could! You have to sort them. Usually we put the data into a **frequency table**. The numbers here lie between 20 and 59. You could take groups, each covering ten years.

age 20–29 30–39 40–49 50–59

It is best to use a tally chart. Go through the list of ages, making a tally mark for each one. The final result is as below. There are 9 tally marks in the first group, so the frequency is 9. The other frequencies are 12, 11 and 8.

age	20–29	30–39	40–49	50–59
tallies	HH IIII	HH HH II	HH HH I	HH III
frequency	9	12	11	8

> **Hint:** it is easy to miss out a number. A useful check is to add up the frequencies. For these numbers, the frequencies should add up to 40.

It is now easier to make sense of the collected data. If someone wanted to know how many people in the group were aged 40 or over, we can see that the answer is $11 + 8$, that is, 19 people.

Exercise 10.1

(Mainly revision)

1 Fifty pupils took an exam. Their marks are below. Make a copy of the frequency table and fill it in using tallies.

```
20   19    8   33   48   30   22   42    9   42
13   18   29   31   42   44   47   30   35   50
41    2   44   11   39   42   48   30   28   48
20   24   31   20   10   33   28   41   40   35
37   27   17    5   26   46   37   27   20   29
```

mark	0–10	11–20	21–30	31–40	41–50
tallies					
frequency					

   a How many pupils got more than 30?
   b How many pupils got 20 or fewer?

2 A doctor is interested in the ages of his patients who are overweight. The ages of 33 patients are below. Make a copy of the frequency table and fill it in.

```
48   41   57   83   41   55   59   61   38   48   79
75   77   54   23   47   56   79   50   61   65   54
36   48   63   59   42   28   32   47   71   82   46
```

age	20–29	30–39	40–49	50–59	60–69	70–79	80–89
tallies							
frequency							

   a How many patients were aged 60 or over?
   b How many patients were under 30?

3 Thirty-six American tourists were asked how many nights they had spent in Britain. The following results were obtained. Copy the frequency table and fill it in.

```
11   16    6   21    1   21    5   22   17
27   17   10    3   20    2   18    8   12
21    6    1    5   13    3   17   14   15
10   18    3    1   11   14   12   23    9
```

number of nights	1–7	8–14	15–21	22–28
tallies				
frequency				

How many tourists stayed for more than two weeks?

4 During a survey of car use, 40 people were asked to record the distance to the nearest mile they travelled by car in a given week. These are the results. Copy and complete the frequency table.

```
23   67   76   85   43   94   48   92   46   28
52   70   77   53   41   48   86   73   56   80
70   70   66   62   54   85   32   29   60   58
43   58   78   44   53   74   52   82   78   47
```

number of miles	20–29	30–39	40–49	50–59	60–69	70–79	80–89	90–99
tallies								
frequency								

How many drivers covered 40 or more miles?

5 Fifty university students take part in a quiz. These are their scores. Copy and complete the frequency table.

```
68   24   75   12   68   90   62   44   31   58
73   79   48   72   66   94   71   38   40   75
61   65   70   51   59   44   95   78   63   32
38   60   55   36   77   20   52   83   73   41
33   54   80   40   32   68   79   94   55   35
```

score	0–19	20–39	40–59	60–79	80–99
tallies					
frequency					

What fraction of students got under 40 marks?

Discrete and continuous data

Some data must consist of whole numbers. When you count the number of people in a room, the result must be a whole number. There may be 6 people, or 7 people, but there can't be $6\frac{7}{8}$ people! This data is **discrete**, because the possible numbers are separated from each other. Here are some other examples.

The number of times someone has taken the driving test

(You can't take it $3\frac{2}{3}$ times.)

The number of letters you received today

(You can't receive 2.4 letters.)

Your shoe size

(It could be $7\frac{1}{2}$, but it can't be $7\frac{3}{8}$.)

But Hortense is my half-sister!

The number of brothers and sisters you have

(You can't have $2\frac{1}{4}$ brothers.)

Discrete data usually consists of whole numbers, but the shoe size example shows they don't have to. Shoe sizes include $7\frac{1}{2}$, $8\frac{1}{2}$, and so on. The results are still discrete data, because the sizes are separated from each other. There's no shoe size between $8\frac{1}{2}$ and 9!

But when you measure something like a length, the result can be any value. The length of a stick can be any positive value. The results aren't separated from each other. The data is **continuous**, as it comes from a range of values in which there are no gaps. For example, the stick could be any length between 1 m and 2 m. Even if you measure it as 1.1 m, it could be a little bit more, at 1.101 m, or a little bit less, at 1.099 m.

Here are some examples of continuous data.

The time you take to run 100 m.

(Perhaps it is any time between 13 seconds and 14 seconds.)

The distance you throw a javelin

(Perhaps it is any distance between 10 m and 20 m.)

The weight of a cat

(Perhaps it is any weight between 2 kg and 3 kg.)

The volume of milk obtained from a cow

(Perhaps it is any volume between 8 litres and 10 litres.)

Exercise 10.2

Which of the following would give discrete data, and which continuous?

1 The weight of a parcel
2 The score on a dice
3 The electrical current flowing through a wire
4 The marks in a test, in which every answer is right or wrong
5 The number of children in a family
6 The temperature of a room

You can put continuous data into a frequency table. Suppose you are measuring weights. The frequency table could look like this.

weight (kg)	10–	20–	30–	40–50
frequency	8	23	17	3

In the first interval are the weights which are at least 10 kg, but less than 20 kg. We can't put 19 kg as the top value in this interval because there may be a weight of 19.9 kg or 19.99 kg or . . .
   In the second interval there are the weights which are at least 20 kg, but less than 30 kg.
   If an item weighs *exactly* 20 kg, you put it in the interval 20–.

weight (kg)	10–20	20–30	30–40	40–50
frequency	8	23	17	3

   You may see the frequency table written like this.
The rule is still the same: an item of *exactly* 20 kg is put in the interval 20–30, not the interval 10–20.

Example   The heights, in centimetres, of 20 children are shown below. Use tallies to copy and complete the frequency table.

142.4 151.7 130.0 155.0 133.1 128.7 157.6 133.0 129.3 120.3
153.7 122.3 118.4 116.3 125.2 123.9 126.1 141.4 153.2 117.6

height (cm)	110–	120–	130–	140–	150–160
tallies					
frequency					

Go through the data, making a tally for each entry. Note that the height of 130.0 cm goes in the third interval, not the second. The results are below. There are 3 tallies for heights between 110 cm and 120 cm, so put 3 for the first frequency. The other frequencies are 7, 3, 2 and 5. The full frequency table is below.

height (cm)	110–	120–	130–	140–	150–160
tallies	///	#### //	///	//	####
frequency	3	7	3	2	5

Exercise 10.3

1 Over a period of 30 days the midday temperatures were found. The results, in °F, are below. Copy and complete the frequency table using tallies.

65.5 73.3 70.0 83.5 96.3 80.5 77.4 76.8 73.9 82.0
67.2 65.2 90.8 74.1 79.3 74.7 76.6 70.7 81.4 82.9
82.3 93.2 91.9 90.2 86.3 79.4 74.1 66.5 69.2 79.8

temperature (°F)	60–	70–	80–	90–100
tallies				
frequency				

For how many days was the temperature under 80 °F?

2 A class of 20 students swam 50 m. Their times, in seconds, are below. Copy and complete the frequency table.

45.6 56.8 69.3 46.2 72.0 67.7 55.3 50.8 49.9 77.8
53.8 57.9 44.2 59.7 60.0 73.1 74.9 64.9 47.7 58.3

time (seconds)	40–	50–	60–	70–80
tallies				
frequency				

How many students took 50 seconds or more?

3 The weights of 30 cats are found. The results (in kg) are below. Copy and complete the frequency table.

2.6 2.7 3.8 3.5 3.5 4.1 3.6 2.9 3.0 2.7
3.2 4.3 2.8 2.6 3.1 4.0 2.6 3.6 2.4 3.7
3.1 4.7 3.7 3.0 3.3 4.1 4.4 3.9 2.6 2.4

weight (kg)	2–	3–	4–5
tallies			
frequency			

How many cats were 3 kg or over?

4 One week the pupils in a maths set are asked to time the number of hours they spent on homework. The results (in hours) are below. Copy and complete the frequency table.

4.5 5.8 2.5 4.1 3.0 4.3 5.9 6.3
7.2 8.6 3.5 4.6 4.0 5.0 6.2 5.1
6.7 2.9 4.1 4.0 5.1 5.0 7.2 6.0

time (hours)	2–	3–	4–	5–	6–	7–	8–9
tallies							
frequency							

How many pupils spent 5 hours or more?

5 Geoff suspects that packets of 'Crackles', which are advertised as having 80 grams of crisps, in fact contain less than 80 grams. He buys and weighs 20 packets. The results (in grams) are given below. Copy and complete the frequency table.

78.3   79.2   81.5   82.4   73.8   76.0   88.4   85.2   83.1   79.3
85.8   77.9   80.0   84.2   83.7   83.6   79.3   72.4   79.4   82.4

weight (grams)	70–	75–	80–	85–90
tallies				
frequency				

How many packets were under 80 grams?

Exercise 10.4   Ma1

Do people overestimate or underestimate measurements?
You can collect your own data to answer this question.

1 Ask 20 people to lift the book, and to guess what its weight is. Then weigh it.

2 Hand 20 people the strip of wood, and ask them to estimate its length. Then measure it.

3 Ask 20 people to raise their hand when they think that exactly 1 minute has passed. Use a stopwatch to find how much time has actually passed.

4 For all of these, record your results on a frequency table. Have most of the people you asked overestimated or underestimated?

You will need:
- weighing machine
- strip of wood
- this book
- tape measure
- stopwatch

Scatter graphs

On the whole, tall people are heavier than short people. If you measure the heights and weights of ten people and plot them on a graph, the result might look like this. This is a **scatter graph**.

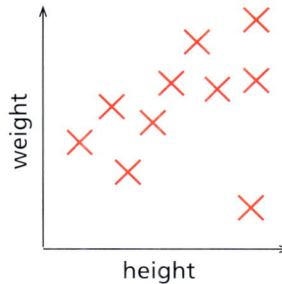

Similarly, if you plot a scatter graph of the engine capacity and top speed of ten cars, the result might look like this. Note that, as the engine capacity increases, the top speed also increases.

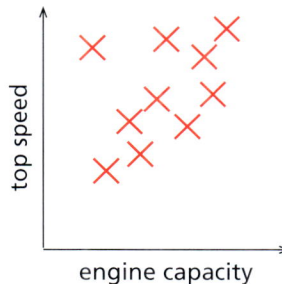

In both cases, there is **correlation** between the two quantities. Because they are both increasing, it is **positive correlation**. As we go from left to right most of the crosses go upwards.

The diagram below shows a scatter graph of the engine capacity and fuel economy of ten cars. There is still correlation, but in the opposite direction. As the engine capacity increases, the number of miles per gallon *decreases*. This is **negative correlation**. As we go from left to right most of the crosses go downwards.

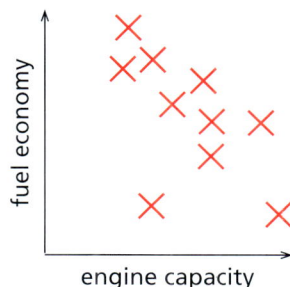

The diagram below shows a scatter graph of the height and maths scores of ten students. As the graph shows, tall people are neither better at maths nor worse – there is no correlation between them at all.

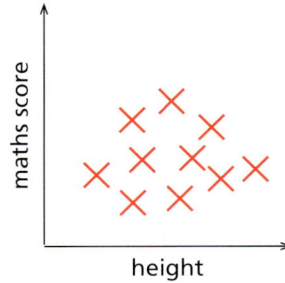

Example   The table below gives the weight and the age of seven babies. Plot a scatter graph on the grid below. What sort of correlation is this?

age (months)	1	3	5	5	6	8	5
weight (kg)	3	5	8	6	7	10	9

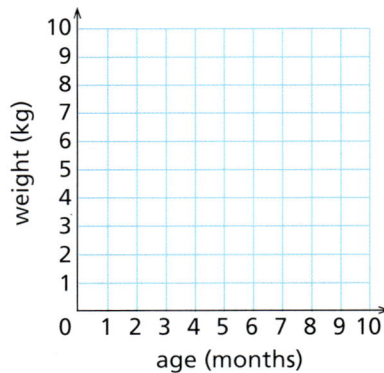

The points on the graph are plotted at (1, 3), (3, 5), and so on. The result is shown below.

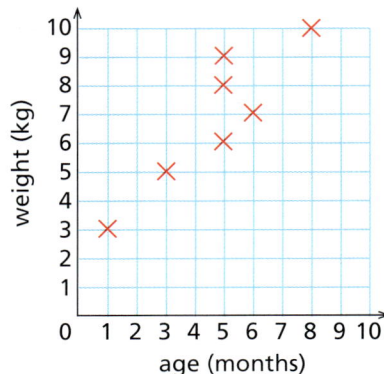

Note that the crosses are going upwards.
There is positive correlation.

Exercise 10.5

1 The scatter graph below shows the wages and ages of several people. What sort of correlation is this?

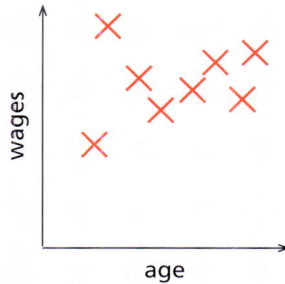

2 Describe the correlation of these three scatter graphs.

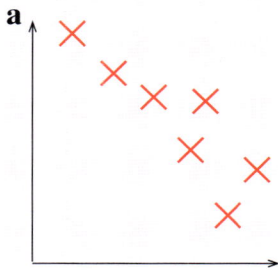

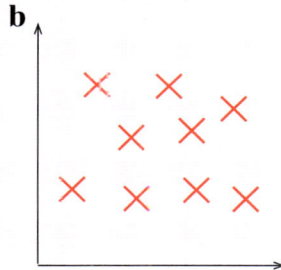

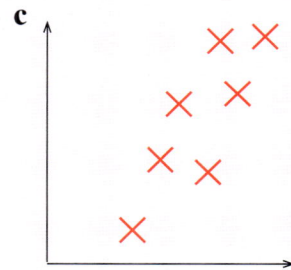

3 Ten students took tests in maths and physics. Their results are below. Using graph paper, make a copy of the grid below and plot a scatter graph on it. What sort of correlation does it show?

maths mark	17	18	10	13	15	15	20	14	13	9
physics mark	14	19	7	10	9	14	18	16	13	6

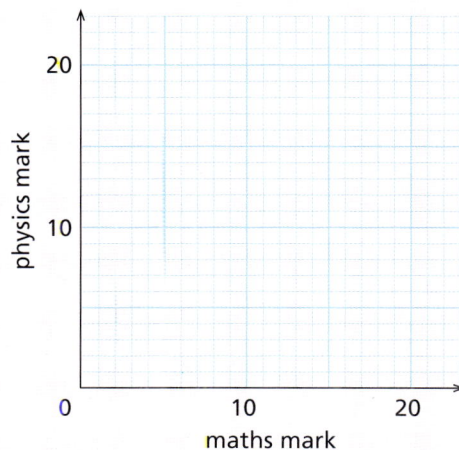

4 Over ten successive Augusts, the rainfall and the average temperature were found. The results are below. Make a copy of the grid and plot a scatter graph on it. What sort of correlation is this?

rainfall (cm)	23	18	3	13	6	10	29	25	18	2
temperature (°C)	15	19	24	20	27	16	16	21	16	23

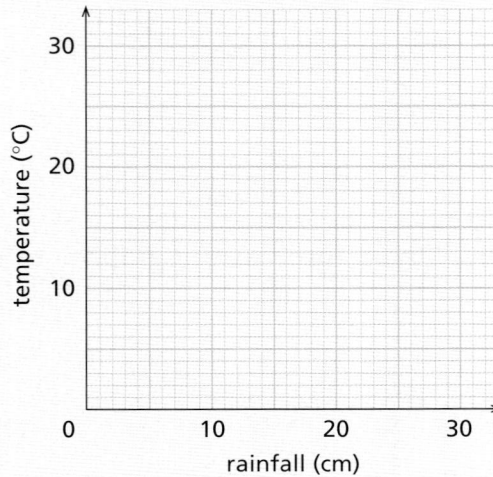

5 Eight students took a test in maths, and then competed in a long jump contest. The results are below. Make a copy of the grid and plot a scatter graph on it. What sort of correlation is this?

maths mark	23	18	3	13	8	10	29	25
long jump distance (m)	1.8	2.9	2.5	3.0	0.6	2.7	2.3	2.1

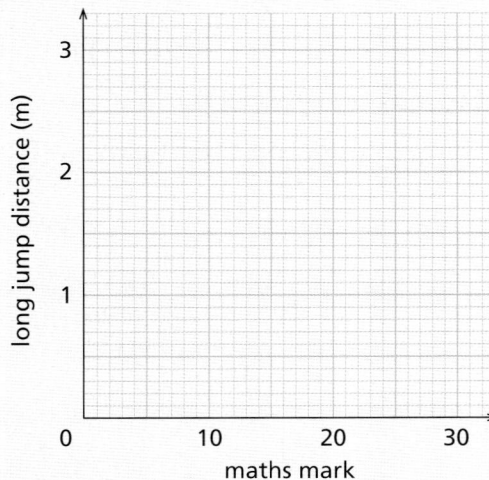

6 The table at the top of page 177 gives the salaries of nine office workers and the distances they could throw a javelin. Make a copy of the grid and plot a scatter graph on it. What sort of correlation is this?

salary (£1000s)	21	19	18	24	16	15	22	20	26
distance (m)	20	21	28	12	10	25	19	20	11

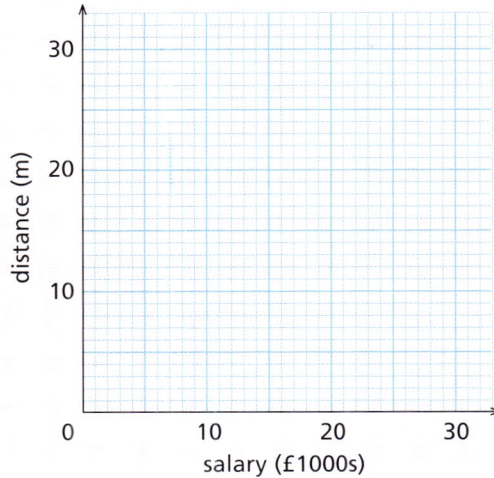

7 Two scatter graphs are shown on the right. Which graph, (i) or (ii), goes with **a** and **b**?

  a The age and time to run 100 m, for ten people aged between 5 and 14

  b The age and time to run 100 m, for ten people aged between 20 and 70

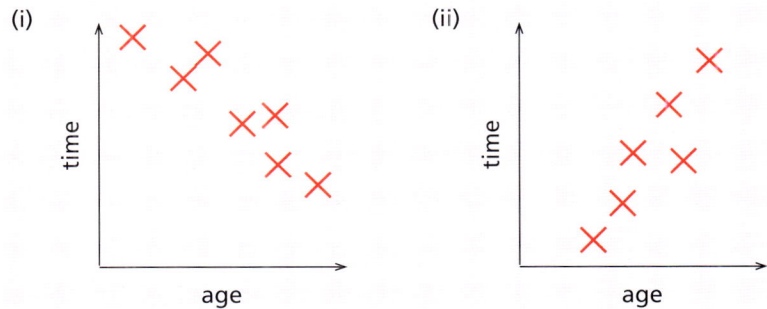

Line of best fit

When a scatter graph shows correlation, we can use it to predict other values. To do this, we need to draw a straight line which is as close as possible to the points. This is a **line of best fit**.

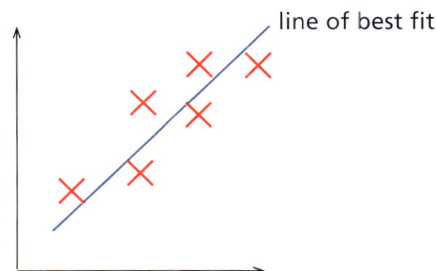

- It must be a *straight* line. Don't join up the points.

This zigzag line is wrong!

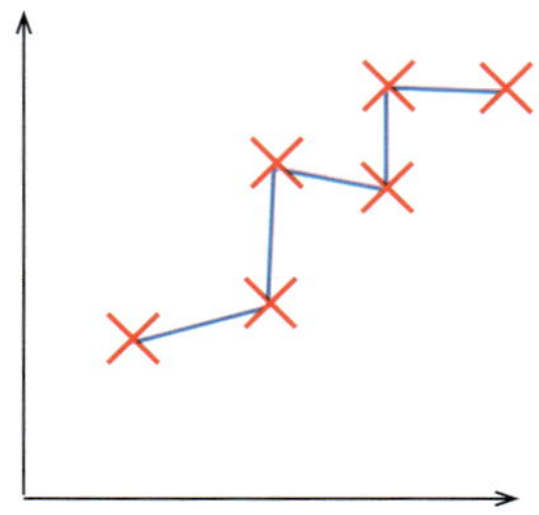

- There should be roughly as many points above the line as below it.

This line is wrong. It's above all the points!

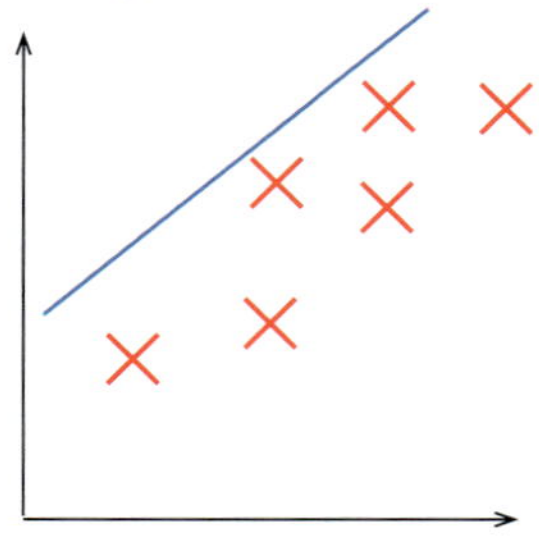

- You are estimating where the line should go. Don't worry if your result is slightly different from someone else's!

From the line of best fit we can make predictions. These are only approximations – if someone else's line is different from yours, your predictions will also be different!

Example   A maths exam has two projects. The results for nine students are given below. Plot a scatter graph for the data, and draw a line of best fit.

first project	17	12	19	9	5	16	11	10	19
second project	17	15	20	10	12	13	11	8	16

Anna scored 13 in the first project, but had to miss the second. What do you think she would have got?

Hita missed the first project, but got 9 in the second. What might she have got in the first?

Plot points at (17, 17) and so on. Draw a line which is as close as possible to the points, as in the graph that follows. Note that there are four points above the line and five below.

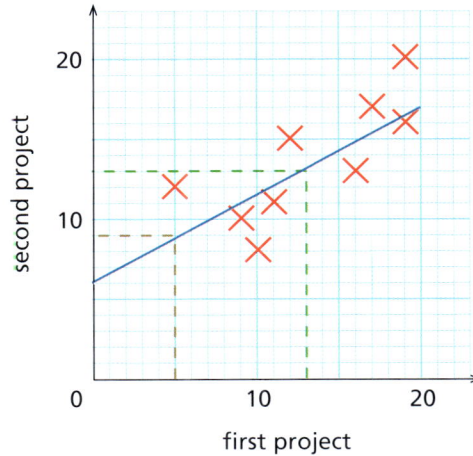

For Anna, find 13 on the *x*-axis, go up to the line and across to the *y*-axis. This is shown by a green dotted line. Read off the *y*-value, 13.

Anna might have got 13 in the second project.

For Hita, find 9 on the *y*-axis, go across to the line and down to the *x*-axis. This is shown by a brown dotted line. Read off the *x*-value, 5.

Hita might have got 5 in the first project.

Exercise 10.6

1 The table below gives the heights of eight girls, all of the same age, and their mothers. Using graph paper, make a copy of the grid below. Plot a scatter graph on it and draw a line of best fit.

height of girl (cm)	132	128	146	141	148	122	158	126
height of mother (cm)	172	159	166	160	177	149	170	153

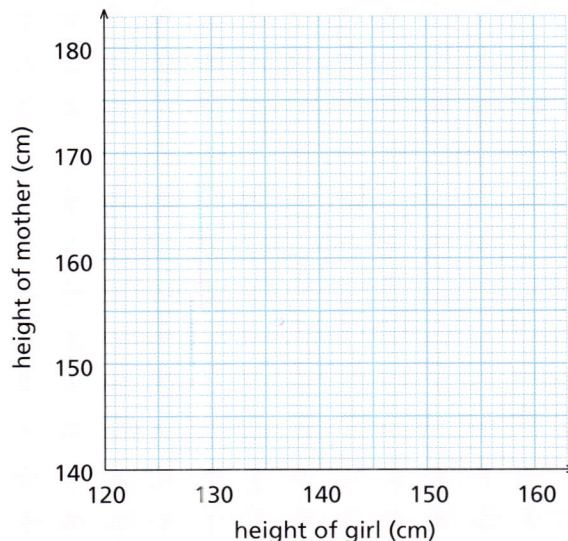

a Another girl of the same age is 140 cm tall. What do you think is her mother's height?

b Another girl of the same age has a mother who is 163 cm tall. What do you think is the girl's height?

2 At seven places across the country, the temperature was measured and the number of hours of sunshine recorded. The results are below. Plot a scatter graph on a copy of the grid and draw a line of best fit.

temperature (°C)	23	20	26	19	18	22	25
hours of sunshine	12	10	14	10	6	13	14

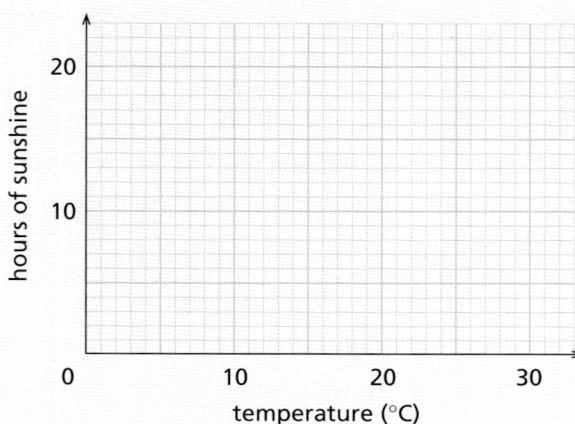

a At another place, the temperature was 23 °C. How many hours of sunshine do you think it had?

b Another place enjoyed 11 hours of sunshine. What do you think the temperature was?

3 The table below gives the total sales of eight pop singles, and the number of weeks they spent in the top 40 chart. Plot a scatter graph on a copy of the grid, and draw a line of best fit.

sales (thousands)	48	59	80	41	77	79	58	66
number of weeks	4	6	10	3	9	8	6	4

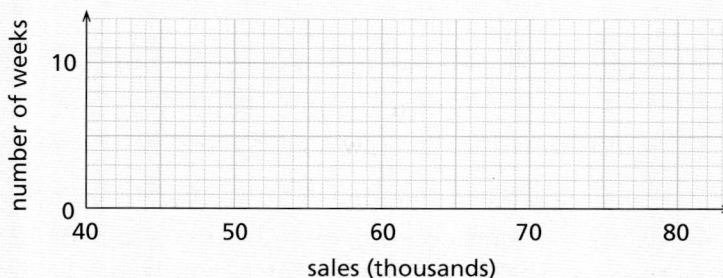

a A single spent 5 weeks in the top 40 chart. What do you think were its sales?

b A single sold 70 000 copies. How many weeks do you think it spent in the chart?

4 Brian has an inaccurate weighing machine. On the machine he weighs eight objects whose weight he knows exactly. The results are below. Plot a scatter graph on a copy of the grid.

exact weight (kg)	0.5	1.2	1.7	1.9	2.1	2.3	2.7	3.0
recorded weight (kg)	0.4	1.1	1.7	1.7	2.2	2.1	2.5	2.7

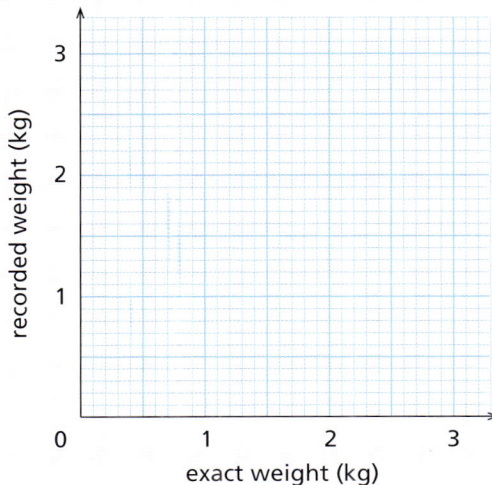

Another object has a recorded weight of 2.3 kg. What do you think its exact weight is?

Cause and effect

Look again at the conversation at the beginning of this chapter. Sean and Jacqui have collected data about people's shoe sizes and the length they could swim. Now you know how to describe what they have found – there is *positive correlation* between these quantities. But does one cause the other?

- Jacqui says that having big feet causes you to be able to swim further.
- Sean says that lots of swimming causes your feet to grow larger.
- There may be other possibilities. For example, as you grow older, your feet grow and you are able to swim further. So both quantities, shoe size and swimming distance, are affected by the third quantity of how old you are.

So, if there is correlation between two quantities, it is possible that change in one quantity causes the other to change. But it is also possible that the change comes from a third quantity, which affects the other two. If there is correlation, don't jump to the conclusion that one quantity causes the other.

Exercise 10.7

For each of the following guess whether there will be positive, negative or no correlation between the quantities. In each case, say whether you think one quantity causes the other, or whether they are both caused by something else.

1 The number of hours spent doing maths homework and the score in the maths exam
2 The score in the maths exam and the score in the science exam
3 The rainfall in August and the number of ice creams sold
4 The attendance at a soccer match and the total incomes of the players on the field
5 The time a student takes to run 100 m and his or her age
6 The time a student takes to run 100 m and the number of words he or she knows

SUMMARY

- Data can be presented in a **frequency table**.
- **Discrete data** take values which are separated from each other. **Continuous data** take values which come from a continuous range of values.
- Continuous data can be presented in a frequency table. If two intervals are 10– and 20–, then the first interval contains all the values which are at least 10 but less than 20. An item of exactly 20 is put in the second interval.
- If pairs of values are plotted, the result is a **scatter graph**. If the points lie roughly on a line, there is **correlation**. If the line goes upwards there is **positive correlation**, and if the line goes downwards there is **negative correlation**.
- A line which is as close as possible to the points of a scatter graph is a **line of best fit**. The line can be used to predict other values. These predictions are only approximations.
- If two quantities are correlated, it does not follow that change in one quantity causes change in the other.

Exercise 10A

1 The numbers below are the prices (in £) of secondhand cars advertised in a local newspaper. Copy and fill in the frequency table, using tallies.

1499	2100	900	850	1900	2500	2100	3200	3800	1999
500	4600	3300	499	2500	4600	3800	3100	4300	1590
2300	2350	1200	3900	800	2540	3900	4900	800	2999
2500	3100	4400	2700	2300	3300	3999	4999	1600	550

price (£)	0–1000	1001–2000	2001–3000	3001–4000	4001–5000
frequency					

How many cars cost over £2000?

2 Which of the following gives discrete data and which continuous?
  a The weight of the apples produced by a tree
  b The number of apples produced by a tree
3 Below are the weights (in kg) of 20 people. Copy and complete the frequency table.

76.3	66.2	57.3	65.9	54.7	66.2	70.0	88.4	84.1	79.2
52.0	61.4	77.4	54.8	83.3	78.0	63.8	77.4	80.7	85.7

weight (kg)	50–	60–	70–	80–90
frequency				

How many people weighed under 70 kg?

4 What sort of correlation is shown in this scatter graph?

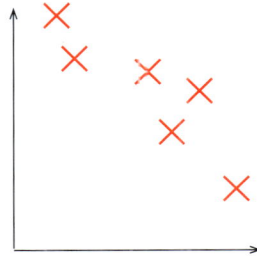

5 For each of the following, say whether you think there will be positive correlation, negative correlation or no correlation.
  a The cost of a car and the number part of its registration number
  b The hours of sunshine and the sales of umbrellas
  c The shoe size and glove size of people
6 The table below gives the marks of eight students in French and German. Using graph paper, make a copy of the grid below. Plot points on the grid to make a scatter graph. What sort of correlation is shown?

French mark	13	6	8	15	28	20	17	12
German mark	6	14	22	20	29	23	20	16

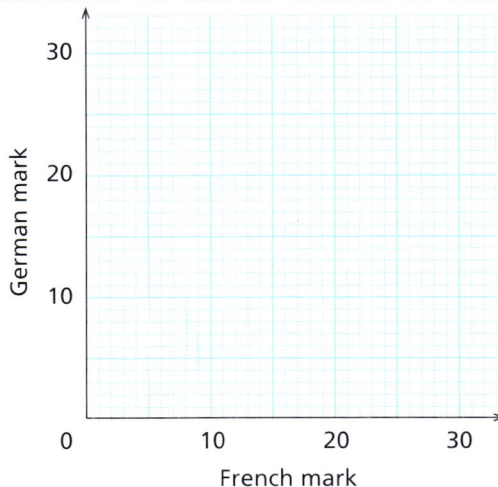

7 Draw a line of best fit on the scatter graph of question 6.
8 Amanda got 23 in the French test (question 6), but was ill for the German test. What do you think she would have got?
9 Boris got 10 in the German test (question 6), but had to miss the French test. What do you think he would have got?
10 Suppose you asked several fellow students for their ages and the amount of pocket money they received. What sort of correlation would you expect to find? Does one cause the other?

Exercise 10B

1 The numbers below give the salaries (in £1000s) of the 20 employees of a small firm. Copy and fill in the frequency table, using tallies.

19	12	13	16	23	27	25	18	13	15
21	20	16	20	26	24	14	17	15	12

salary (£1000s)	10–15	16–20	21–25	26–30
frequency				

How many people earned under £26 000?
2 Which of the following gives discrete data and which continuous?
   a The number of birthday presents you receive
   b The weight of the birthday presents you receive
3 Thirty athletes run a 100 m race. Their times, in seconds, are given below. Copy and complete the frequency table.

11.23	11.25	10.65	10.55	11.03	11.31	11.62	10.66	10.52	10.70
11.71	11.31	11.94	11.51	10.63	10.74	11.04	10.31	10.60	11.52
10.31	10.44	10.82	10.42	11.57	11.63	10.22	10.42	11.80	10.97

time (seconds)	10.0–	10.5–	11.0–	11.5–12.0
frequency				

How many athletes took 10.5 seconds or longer?
4 What sort of correlation is shown in this scatter graph?

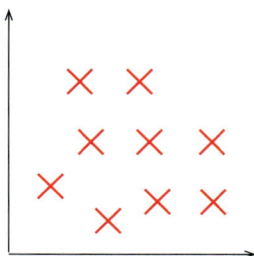

5 Three scatter graphs are shown below. Match up the graphs with **a**, **b** or **c** below.

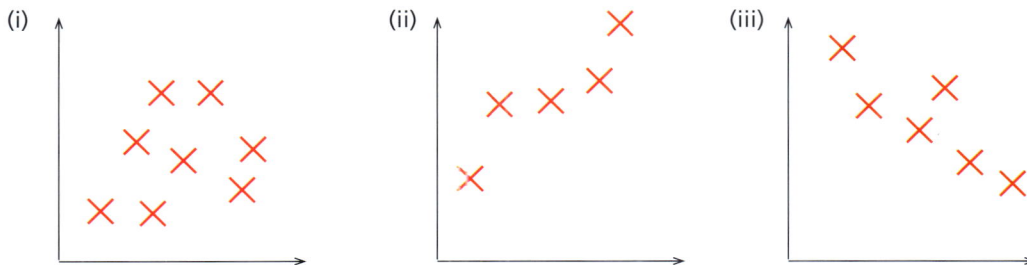

(i)

(ii)

(iii)

  a The earnings of a professional tennis player and her place in the world rankings
  b The attendance at a one-day cricket match and the number of runs scored
  c The earnings of a professional snooker player and the number of tournaments he has won

6 The table below gives the ages of eight young children and the average time they spend asleep each day. Make a copy of the grid, and plot a scatter graph on it. What sort of correlation is shown?

age (years)	0.5	1.0	1.8	2.3	3.0	3.2	3.5	3.8
time asleep (hours)	18	12	10	11	8	11	9	8

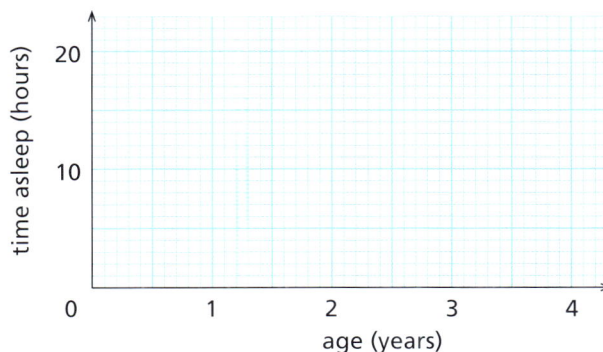

7 Draw a line of best fit on your scatter graph for question 6.
8 Ian is 1.5 years old. How long do you think he spends asleep?
9 Nancy is a young child who sleeps for 14 hours a day. How old do you think she is?
10 Suppose you investigated several pizza shops, and in each one found the cost of their cheapest pizza and their cheapest cola. What sort of correlation would you expect to find? Does one cause the other?

Exercise 10C Ma1

If you have a scientific calculator, it probably has a button which gives a random number between 0 and 1. It may be labelled RAN#. Use it to find 60 random numbers, and record your results in a copy of the frequency table below. Are the numbers evenly spread between the intervals?

random number	0–	0.2–	0.4–	0.6–	0.8–1
frequency					

Exercise 10D Ma1

There are many investigations you can do about correlation. Some suggestions are given below. Collect data for one of these, and plot a scatter graph for your findings. Plot a line of best fit, and use it to make predictions. Is there a connection of cause and effect between the quantities?

- The heights of the wife and husband in ten couples
- The price and the average daily sales of eight newspapers
- The position in the league of twelve football teams and the number of goals they have scored so far in the season

11 Transformations and congruence

Remember chapter 8.

Transformations move points and figures. You have already met one sort of transformation: enlargements. Here we deal with two other sorts: translations and reflections.

Translations

Mathematical translations have nothing to do with translating English into Spanish! In mathematics, a **translation** is the simplest sort of transformation. A translation moves figures without altering their shape, size or the direction in which they are pointing. You do a translation when you pick up a chair and move it, without turning it round or putting it on its side.

It is easy to describe a translation. You say how far you have moved left or right and how far you have moved up or down.

Example   The grid below has lines which are 1 cm apart. Translate the dot 3 cm to the right and 2 cm upwards.

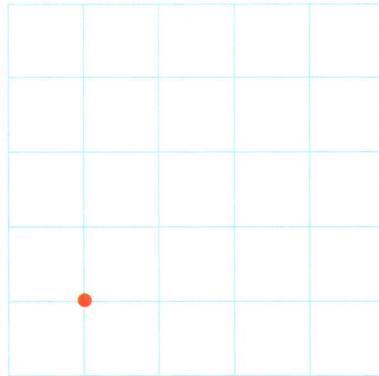

Move the dot 3 cm to the right and then 2 cm upwards. The result is shown below.

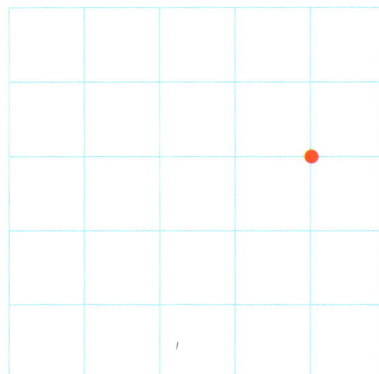

Note. We did the right movement first and then the upward movement. It wouldn't make any difference if we did the upwards movement first!

Exercise 11.1

Copy this diagram onto squared paper and move the dots by the translations given.

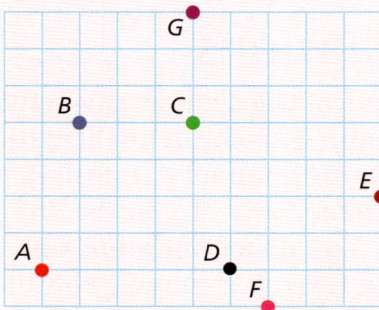

1 Dot *A* by 5 squares to the right and 1 square upwards
2 Dot *B* by 3 squares to the right and 2 squares upwards
3 Dot *C* by 1 square to the left and 3 squares upwards
4 Dot *D* by 3 squares to the right
5 Dot *E* by 4 squares upwards
6 Dot *F* by 2 squares to the right and 3 squares upwards
7 Dot *G* by 4 squares to the left and 2 squares downwards

The same procedure holds for more complicated figures. Every part of the figure moves by the same amount, and in the same direction.

Example   Move the triangle *ABC* by 3 cm to the left and 2 cm upwards.

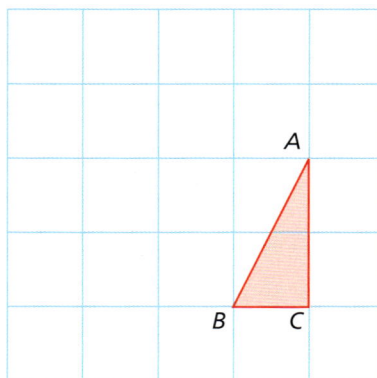

Move point *A* 3 cm to the left and 2 cm upwards.
Move point *B* 3 cm to the left and 2 cm upwards.
Move point *C* 3 cm to the left and 2 cm upwards.

Now join up the sides of the triangle. The result is shown below. The whole triangle has moved 3 cm to the left and 2 cm upwards.

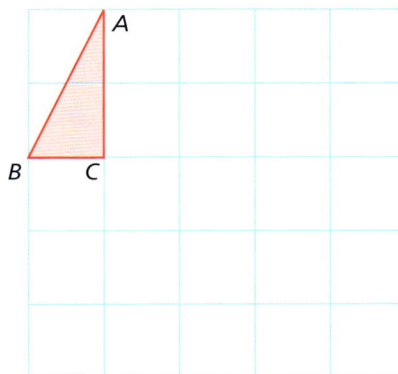

Exercise 11.2

Copy both the diagrams onto squared paper and move the shapes by the translations given.

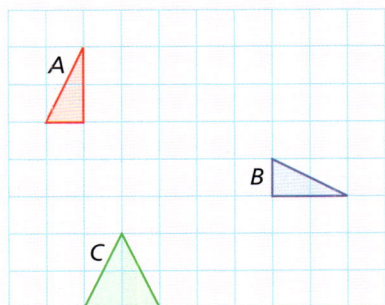

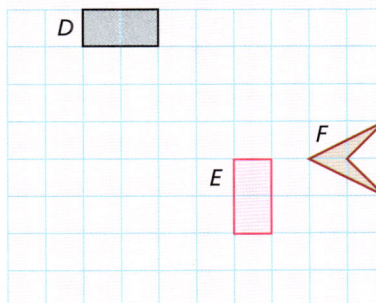

1 Triangle *A* by 4 to the right and 1 upwards
2 Triangle *B* by 5 to the left and 3 upwards
3 Triangle *C* by 3 to the right
4 Rectangle *D* by 3 to the right and 2 downwards
5 Rectangle *E* by 4 to the left
6 The 'arrowhead' *F* by 3 to the left and 3 downwards

Exercise 11.3 (Ma1)

This is a game involving translations. Two people play it, while an umpire keeps track of the moves. A coin is placed in the middle of the board, as shown. The two players have their backs to the board, and take turns to call out a translation of the coin, which the umpire then moves. The loser is the person who sends the coin off the edge of the board!

You will need:
- chess board
- coin

You can play safe by trying to keep to the centre of the board, or be bold and move as close to the edge as you dare!

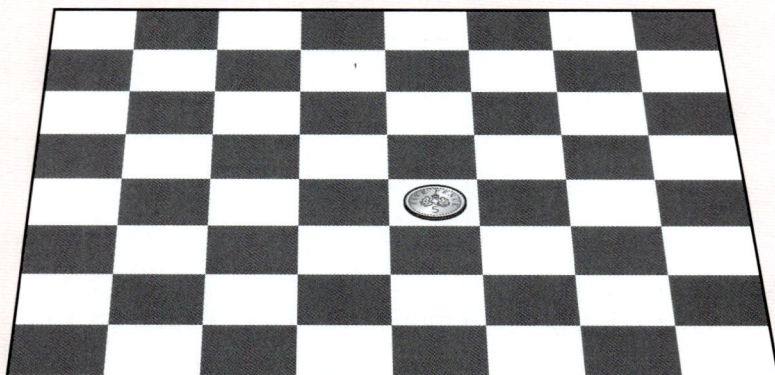

Use of coordinates

Translations are made easier when the points and shapes are on a coordinate grid. The translation is done by ordinary arithmetic!

Example   The dot on this coordinate grid is at (3, 1). It is translated 2 to the right and 3 up. What are its new coordinates?

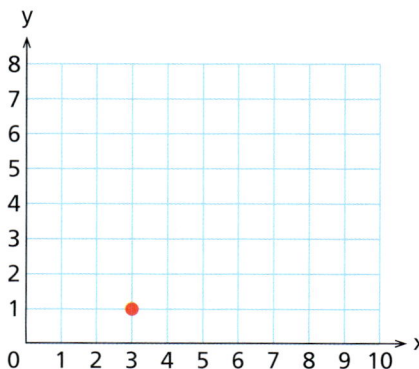

Move the point 2 right and 3 up, as shown.

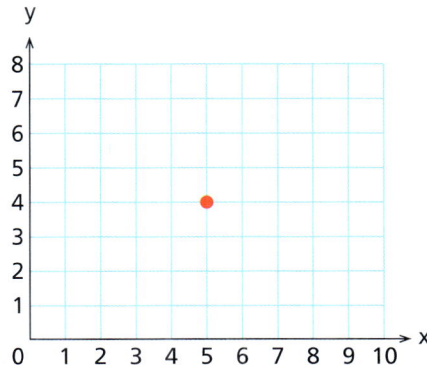

The new coordinates are (5, 4).

Exercise 11.4

Make a copy of this diagram on squared paper, and translate point *A* as described. In each case write down the new coordinates of *A*.

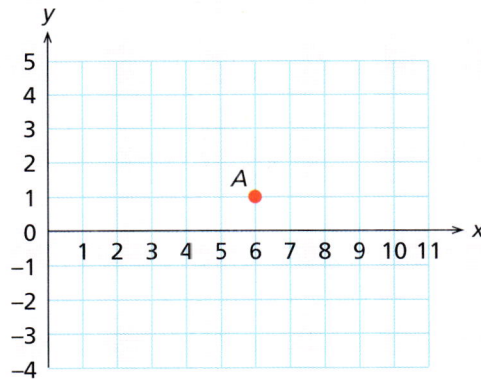

1 3 to the right and 1 up
2 4 to the right and 2 up
3 1 to the right and 2 down
4 4 to the left and 3 up
5 5 to the left and 2 down

You should have noticed that a translation just adds to or subtracts from the coordinates.

To move 3 to the right, add 3 to the *x*-coordinate.
To move 3 to the left, subtract 3 from the *x*-coordinate.
To move 3 upwards, add 3 to the *y*-coordinate.
To move 3 downwards, subtract 3 from the *y*-coordinate.

Example   Suppose *A* is at (4, 5). What are its coordinates after it has been translated 3 to the right and 2 downwards?

You don't need to draw a picture. Add 3 to the *x*-coordinate, and subtract 2 from the *y*-coordinate.
The new coordinates are (7, 3).

Exercise 11.5

In these questions write down what the coordinates of the point will be after the translation. Do it by arithmetic – don't draw a diagram!

 1 Point at (5, 8); translate 4 to the right and 7 upwards
 2 Point at (3, 9); translate 6 to the right and 9 upwards
 3 Point at (4, 1); translate 2 to the left and 2 upwards
 4 Point at (5, 8); translate 8 to the right and 5 downwards
 5 Point at (4, 3); translate 5 to the left and 6 downwards
 6 Point at (−3, 4); translate 7 to the right and 2 upwards
 7 Point at (4, −6); translate 6 to the right and 3 upwards
 8 Point at (−8, −2); translate 4 to the left and 8 upwards
 9 Point at (−5, 0); translate 6 to the left and 4 downwards
10 Point at (0, −9); translate 5 to the right and 7 downwards

Finding the translation

Look at the two triangles in the diagram, one red and one blue. They are the same shape and size, and they are pointing in the same direction. The blue triangle is a translation of the red triangle. What is the translation?

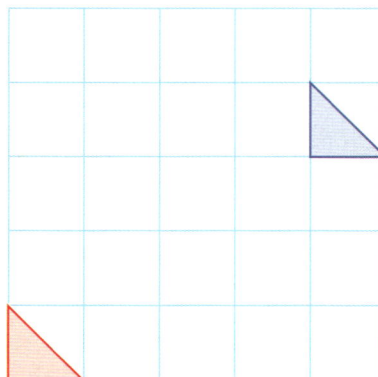

Look at the points of the triangles.
The blue top point is 4 to the right and 3 above the red top point.

The blue bottom point is 4 to the right and 3 above the red bottom point.

The blue left point is 4 to the right and 3 above the red left point.

All points of the red triangle have been moved in the same way, 4 to the right and 3 up. This is the translation.

Exercise 11.6

1 For **a** to **e** find the translation which has taken the red point to the blue point.

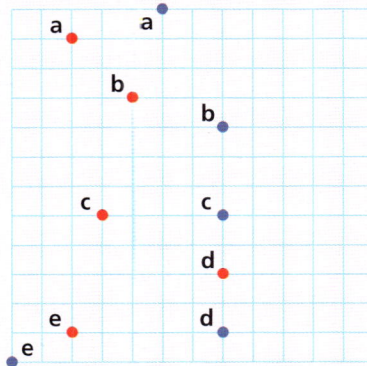

2 For **a** to **e** find the translation which has taken the red triangle to the blue triangle.

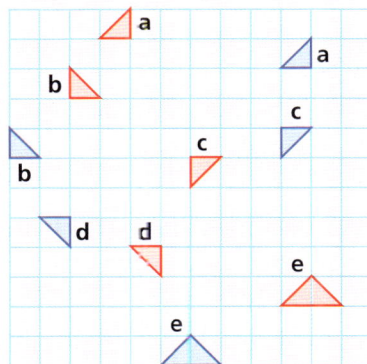

If the points are given as coordinates, we can find the translation immediately. The change in *x*-values gives the movement to the right or left, and the change in *y*-values gives the movement upwards or downwards.

> If the x-coordinate increases, it is a translation to the right.
> If the x-coordinate decreases, it is a translation to the left.
> If the y-coordinate increases, it is a translation upwards.
> If the y-coordinate decreases, it is a translation downwards.

Example A translation moves (3, 7) to (5, 1). What is the translation?

We don't have to draw a diagram! The *x*-coordinate has increased by 2, and the *y*-coordinate has decreased by 6.
    The translation is 2 to the right and 6 downwards.

Exercise 11.7

For questions 1 to 6, describe the translation which has moved the point as given.

1 (1, 1) to (4, 6)
2 (3, 6) to (8, 10)
3 (6, 2) to (2, 5)
4 (3, 4) to (8, 0)
5 (4, 9) to (1, 2)
6 (4, 6) to (−2, −5)
7 A translation moves (1, 1) to (2, 7). Where does this translation take (3, 4)?
8 A translation moves (7, 3) to (3, 1). Where does this translation take (6, 5)?
9 A translation moves (−2, −3) to (5, 6). Where does this translation take (2, 7)?
10 A translation moves (4, 2) to (−2, −8). Where does this translation take (6, −4)?

Vectors

It is a bit long-winded to keep saying: 'translate 3 to the right and 4 upwards'. It would be good to have a shorter way of describing translations.

The important things about a translation are the two distances moved, horizontally and vertically. Can you think of a way to show these two numbers?

The standard way to display the distances moved is to use a **vector**:

$$\binom{3}{4}$$

This vector describes a translation of 3 to the right and 4 upwards.

Note that the distance moved horizontally (along the *x*-direction) must be on top, and the distance moved vertically (along the *y*-direction) must be on the bottom.

If the translation only moves points horizontally, then the *y* change is 0. So there will be a 0 for the bottom number. This vector moves points 3 units to the right.

$$\binom{3}{0}$$

In vectors:
- the top number shows horizontal movement
- the bottom number shows vertical movement.

Exercise 11.8

For questions 1 to 5 write down the vector corresponding to the translation.

1 2 to the right and 5 upwards
2 5 to the right and 3 upwards
3 4 to the right and 7 upwards
4 7 to the right
5 6 upwards

For questions 6 to 10 write the translation corresponding to the vector.

6 $\begin{pmatrix} 2 \\ 4 \end{pmatrix}$ **9** $\begin{pmatrix} 3 \\ 0 \end{pmatrix}$

7 $\begin{pmatrix} 3 \\ 8 \end{pmatrix}$ **10** $\begin{pmatrix} 0 \\ 9 \end{pmatrix}$

8 $\begin{pmatrix} 9 \\ 3 \end{pmatrix}$

11 A translation takes (3, 6) to (4, 9). Write down the vector for this translation.
12 A translation takes (2, 7) to (10, 9). Write down the vector for this translation.
13 Where does the vector of question 6 take (2, 5)?
14 Where does the vector of question 7 take (5, 8)?

What about moving to the left, or moving downwards? Here the usefulness of the vector form shows itself. We can use negative numbers!

- *A movement to the left is a negative movement.*
- *A movement downwards is a negative movement.*

Examples   Write down a vector corresponding to a movement of 3 to the left and 2 downwards.

The movement of 3 to the left is a movement of -3.
   The movement of 2 downwards is a movement of -2.
   Write this as a vector, as below.

$$\begin{pmatrix} -3 \\ -2 \end{pmatrix}$$

What translation does this vector correspond to?

$$\begin{pmatrix} 4 \\ -3 \end{pmatrix}$$

The top number 4 is positive. So it represents a move of 4 to the right.
  The bottom number −3 is negative. So it represents a move of 3 downwards.
  The translation is 4 to the right and 3 downwards.

How useful negative numbers are! If we didn't have negative numbers there would be four different types of translation:

to the right and up
to the right and down
to the left and up
to the left and down

All these four cases can be dealt with at once, by letting the negative sign tell us whether it is to the right or left, and whether it is up or down!

Exercise 11.9

For questions 1 to 5 write down the vector for the translation given.

1 4 to the left and 6 downwards
2 6 to the left and 3 upwards
3 7 to the right and 1 downwards
4 7 to the left
5 8 downwards

For questions 6 to 10 write the translation corresponding to the given vector.

6 $\begin{pmatrix} -4 \\ 7 \end{pmatrix}$

7 $\begin{pmatrix} 3 \\ -8 \end{pmatrix}$

8 $\begin{pmatrix} -2 \\ -5 \end{pmatrix}$

9 $\begin{pmatrix} -14 \\ 0 \end{pmatrix}$

10 $\begin{pmatrix} 0 \\ 12 \end{pmatrix}$

Exercise 11.10 {Ma1}

Do you play draughts or chess? In both these games you move pieces about a board, that is, you *translate* them. You could write these translations as vectors – for example, a pawn in chess can move one step forwards. Its vector movement is

$\begin{pmatrix} 0 \\ 1 \end{pmatrix}$

Find the vectors of the possible moves of the other pieces.

Exercise 11.11 {Ma1}

Chess and draughts are played on boards with squares. Some games have boards covered by hexagons, as shown below. Can you think up a way to describe moves on a hexagonal board?

Remember:

You can tessellate with equilateral triangles, with squares or with regular hexagons.

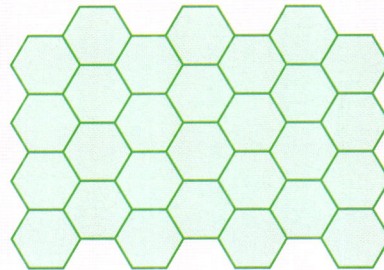

Reflections

In Book 1 we looked at the **symmetry** of various figures. A figure has a line of symmetry if a fold line cuts it in half exactly so that the bit on one side of the line matches the bit on the other side.

Look at this shape – if it was folded along the dotted line, the bits on either side of the line would match exactly.

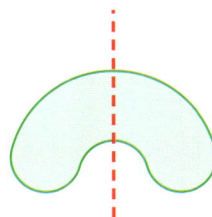

Exercise 11.12

(Mainly revision)

Copy each of these figures, and mark the lines of symmetry with dotted lines.

1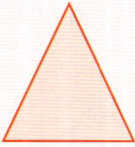

2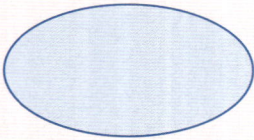

3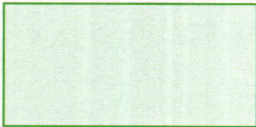

4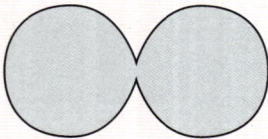

5

In fact, given *any* shape, you can extend it so that it has a line of symmetry. If you draw a shape (in ink or paint) on a piece of paper, then fold the paper while the ink is still wet, the shape will be symmetrical about the fold line.

A Swiss psychiatrist called Rorschach asked his patients to say what they saw in ink blot patterns like this. He analysed their mental state from their answers! The patterns are sometimes called Rorschach blots.

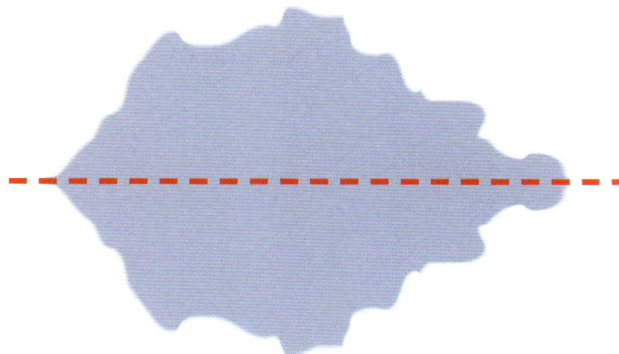

The operation of finding the image on the other side of the shape is **reflection**. It is like finding the reflection of the shape in a mirror. The fold line is often called the **mirror line**.

Exercise 11.13

Draw a shape on a piece of paper with water-based ink. Before the ink dries fold the paper over and press firmly. Convince yourself that the shape is now symmetrical about the line of the crease. Do you see anything in the shape?

> **Remember:**
>
> *Translations and reflections are both types of transformation.*

Doing reflections on grids

When a point is reflected in a line, it moves to the same distance on the other side of the line. It is rather inaccurate to do the measurement by a ruler, so it is much easier and more accurate if it is done on a grid. Then the distance can be found by counting squares.

Example   The point *P* is to be reflected in the vertical dotted line shown. Draw the image of point *P*.

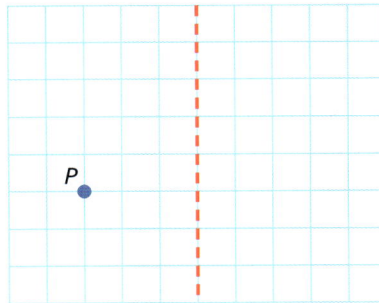

By counting squares, we can see that *P* is 3 units to the *left* of the dotted line. Its image will then be 3 units to the *right* of the dotted line. Label this image as *P'*. The result is shown below.

We can use this method to reflect whole shapes.

Example   Reflect the triangle in the horizontal dotted line.

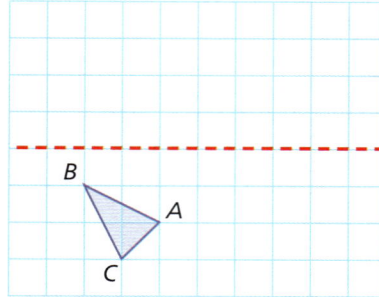

Each of the vertices has to be reflected.

The point A is 2 units below the line, so it moves to A' which is 2 units above.

The point B is 1 unit below the line, so it moves to B' which is 1 unit above.

The point C is 3 units below the line, so it moves to C' which is 3 units above.

Join up the points to form triangle $A'B'C'$.

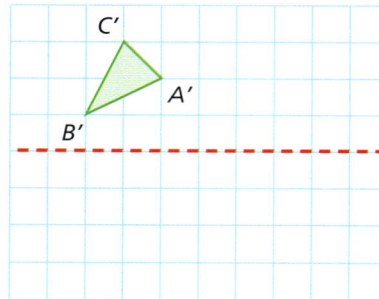

Exercise 11.14

1 Make a copy of the diagram on the right on squared paper, and reflect the points in the dotted line. Label the reflected points A', B', C'.

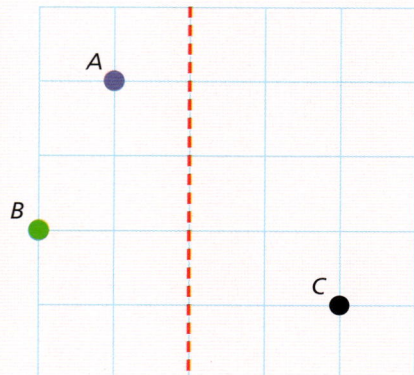

2 Make a copy of the diagram below, and reflect the points in the dotted line. Label the reflected points D', E'.

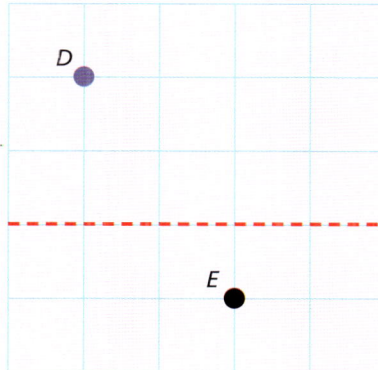

3 Make a copy of the diagram below, and reflect the shapes in the dotted lines. Label the new shapes F', G', H'.

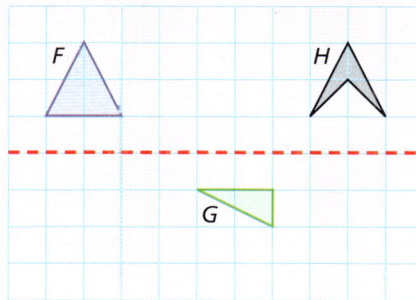

The figures so far have all been on one side of the mirror line. The figure could straddle the mirror line. Follow the same procedure when reflecting – some points will move to the left of the line, and some will move to the right.

Example Reflect the triangle ABC in the dotted line.

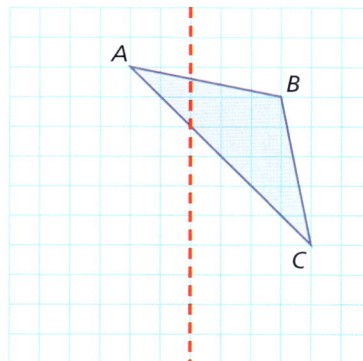

Point *A* is 2 units to the left of the line, so it moves to 2 units to the right of the line at *A'*.

Point *B* is 3 units to the right, so it moves to 3 units to the left.

Point *C* is 4 units to the right, so it moves to 4 units to the left.

Join up the points of the new triangle. The result is shown below.

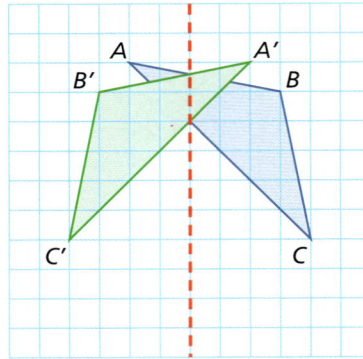

The original triangle and its reflection now form a different shape. As before, the mirror line is a line of symmetry of the new shape.

Exercise 11.15

In these questions make copies of the diagrams, and reflect the shapes in the dotted lines.

1

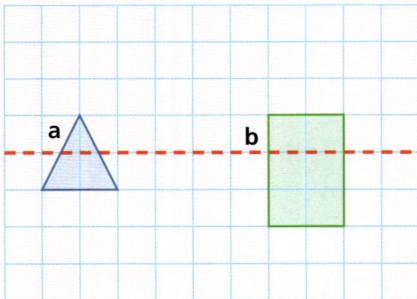

2

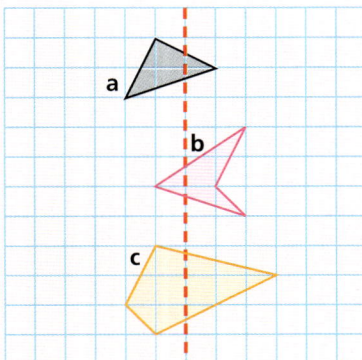

Slanting mirror lines

Things have been made easy so far by having mirror lines which were either vertical or horizontal. Then you only had to count the squares to the right or left, or up or down. If the mirror line is slanting then it becomes trickier.

Example   Reflect point *A* in the slanting mirror line.

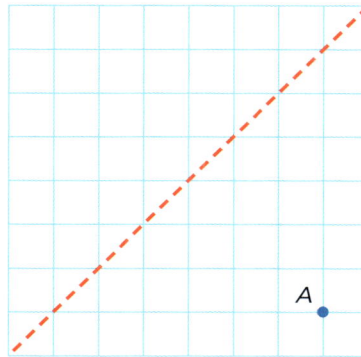

The mirror line in this diagram is sloping at 45°. We can still do reflections by counting, but by going diagonally across the squares. The point *A* is 3 diagonals below right of the mirror line. Count out 3 diagonals above left of the line. This takes *A* to its **mirror image**, *A'*.

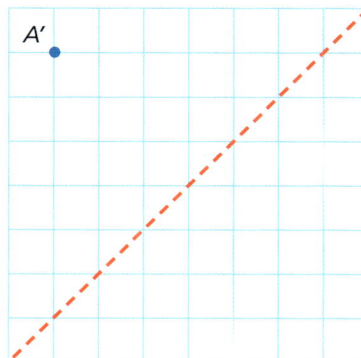

Exercise 11.16

In these questions make copies of the diagrams and reflect the points in the slanting mirror lines.

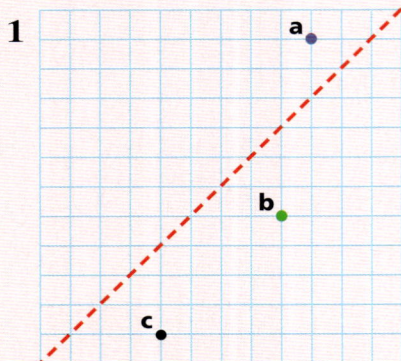

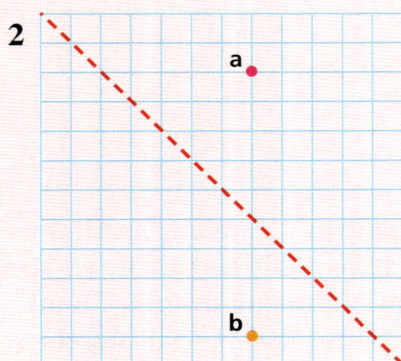

1

2

Reflections by coordinates

It helps to do reflections if the figures are placed on a grid. It helps even more if the grid is numbered, so that points have coordinates. Let us take a coordinate grid, and take the *x*-axis as the mirror line.

Reflect point *A*, at $(3, 2)$, in the *x*-axis.

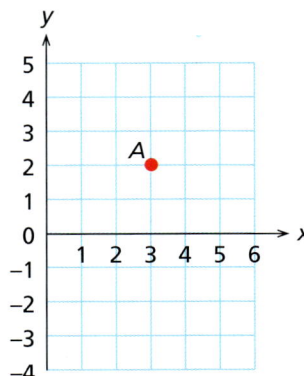

Point A is 2 squares above the x-axis. After reflection, it will be 2 squares below the x-axis, at $(3, -2)$.

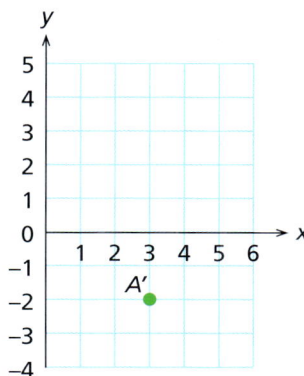

What about B, at $(1, 4)$? This is 4 squares above the x-axis, so after reflection it is 4 squares below, at $(1, -4)$.

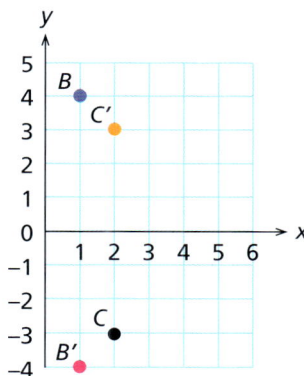

What about C, at $(2, -3)$? This is 3 squares *below* the x-axis, so after reflection it is 3 squares above the x-axis, at $(2, 3)$.

Exercise 11.17

For these questions, draw a coordinate grid on squared paper and reflect the points in the x-axis.

1 $(2, 6)$ **3** $(1, 3)$ **5** $(0, -4)$
2 $(5, 3)$ **4** $(6, -8)$

Can you spot the rule? For all points, a positive y-coordinate becomes negative, and a negative y-coordinate becomes positive. That sounds complicated – we can have just one rule for this. To turn positive to negative and negative to positive, multiply by -1.

● *To reflect a point in the x-axis, multiply its y-coordinate by -1.*

You may be able to guess the rule for reflection in the *y*-axis. Here is an exercise to help you discover it.

Exercise 11.18

For these questions, draw a coordinate grid on squared paper and reflect the points in the *y*-axis.

1 $(3, 2)$ **2** $(5, 3)$ **3** $(-1, 8)$ **4** $(-5, 3)$ **5** $(8, 0)$

Did you get the rule?

● *To reflect in the y-axis, multiply the x-coordinate by −1.*

Remember chapter 6.

The vertical line through $(3, 0)$ on the *x*-axis is the line $x = 3$. The horizontal line through $(0, 2)$ on the *y*-axis is the line $y = 2$. We can reflect points and figures in these lines.

Example   Reflect the point *B*, at $(1, 2)$, in the line $x = 3$.

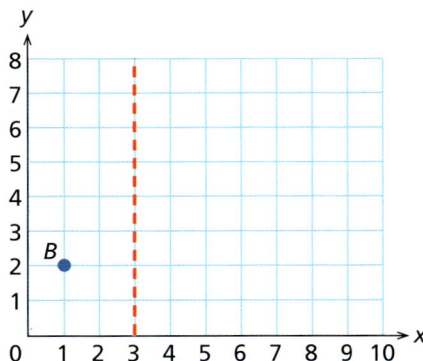

The point *B* is two units to the left of the line $x = 3$. After reflection, it is two units to the right of the line. Hence its *x*-coordinate is 5. Its *y*-coordinate is unchanged.

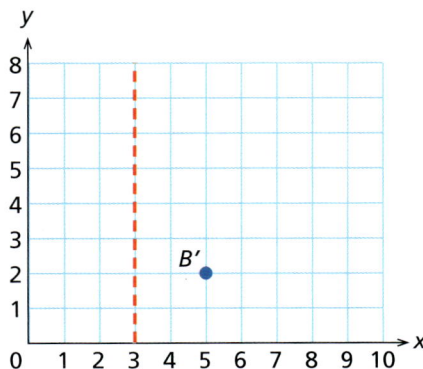

The point is reflected to $(5, 2)$.

Exercise 11.19

For these questions, draw a coordinate grid on squared paper and reflect the point in the line given.

1. $(1, 3)$, in the line $x = 4$
2. $(4, 2)$, in the line $x = 3$
3. $(2, 6)$, in the line $x = 4$
4. $(-4, 3)$, in the line $x = -2$
5. $(5, 1)$, in the line $x = 5$
6. $(3, 2)$, in the line $y = 4$
7. $(4, 7)$, in the line $y = 6$
8. $(4, 3)$, in the line $y = -1$
9. $(7, -2)$, in the line $y = -5$
10. $(7, 1)$, in the line $y = 1$

Reflection in a diagonal line

Now let us take as mirror line the diagonal line which goes through the origin, $(1, 1)$, $(2, 2)$, $(3, 3)$, and so on. For every point (x, y) on this line, $y = x$.

Example   Reflect A, at $(1, 5)$, in the diagonal line $y = x$. This point is 2 diagonal steps away from the mirror line.

Go to 2 diagonal steps on the other side of the line. You will end up at $(5, 1)$.

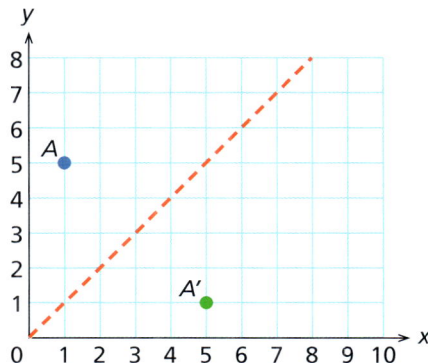

Can you see the rule?

Exercise 11.20

For these questions, draw a coordinate grid on squared paper and reflect the points in the diagonal line, $y = x$.

1 $(2, 1)$ **3** $(3, 6)$ **5** $(6, 6)$
2 $(4, 5)$ **4** $(6, 8)$

Did you get the rule? It is this: swap round the x-coordinate and the y-coordinate.

Congruence

After translating or reflecting a figure, its shape and size are unchanged. If we cut out the original figure, we could fit it exactly over the new figure. The two figures are **congruent** to each other.

Exercise 11.21

You will need:

- graph paper
- pencil
- ruler
- scissors

1 Draw a triangle ABC on graph paper, and translate it by the methods of this chapter. Cut out the second triangle, and show that it will fit exactly on top of the original triangle.
2 Draw a triangle ABC on graph paper, and reflect it by the methods of this chapter. Cut out the second triangle, and show that it will fit exactly on top of the original triangle.
3 Colour your cut-out triangles on one side of the paper. When you fitted the cut-out triangle, what was the difference between question 1 and question 2?

Example   The diagram below shows an isosceles triangle ABC, cut down the middle by AD. Find a pair of congruent triangles.

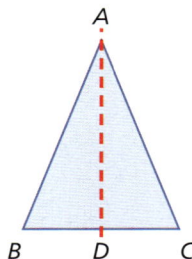

There are two triangles: if we reflect the left-hand triangle ABD in the line AD, we get the right-hand triangle ACD. These two triangles are congruent to each other.

Exercise 11.22

In questions 1 to 5, which of the shapes are congruent to each other?

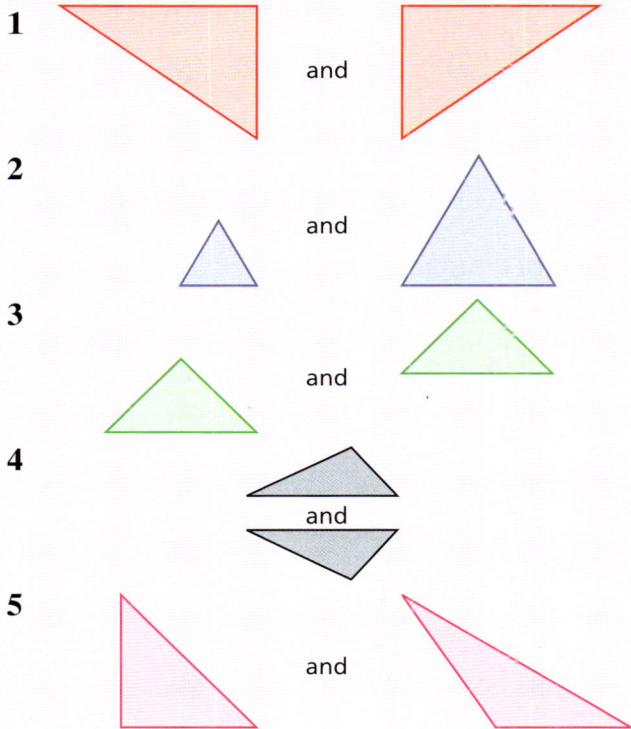

1   and

2   and

3   and

4   and

5   and

6 Which of the shapes below are congruent to each other?

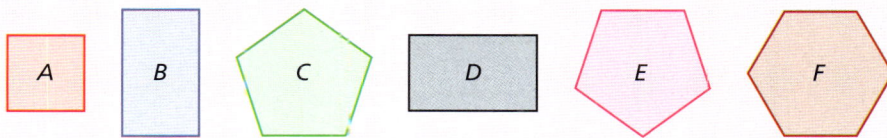

A   B   C   D   E   F

7 A rhombus is a quadrilateral in which all sides are equal. The diagram below shows a rhombus *ABCD*. What are the lines of symmetry of the rhombus? Write down any pairs of congruent triangles.

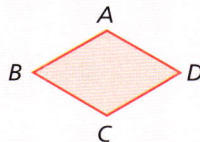

8 The diagram below shows a kite *EFGH*. Write down a pair of congruent triangles.

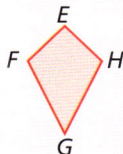

SUMMARY

■ **Translations** and **reflections** are both types of **transformation**.
■ A translation moves all points in a fixed direction. If the coordinates of a point are given, you can just add to or subtract from the coordinates.
■ You can use a **vector** to describe a translation. The top number of the vector is the distance moved horizontally, and the bottom number is the distance moved vertically.
■ By using negative numbers, you can give vectors which describe translations to the left or downwards.
■ A reflection moves points to the other side of a **mirror line**.
■ To reflect points in the *x*- or *y*-axes, multiply their *y*- or *x*-coordinates (respectively) by −1.
■ If two figures have been obtained from each other by translation or reflection, they are **congruent** to each other.

Exercise 11A

1 Make a copy of the diagram below on squared paper.

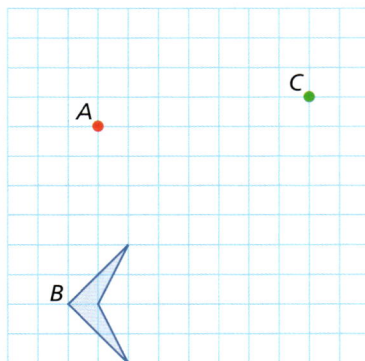

    a Translate the dot *A* two units to the right and three units up.
    b Translate the shape *B* one unit to the right and two units up.
    c Translate the dot *C* two units to the left and three units down.
    d What translation would take dot *A* to dot *C*?
2 Write down the vector corresponding to the translation of question 1a.
3 A translation takes $(3, 1)$ to $(-2, 5)$. Write down the vector corresponding to this translation.
4 A translation is represented by the vector below. Where does the translation take $(-2, 4)$?

$$\begin{pmatrix} 3 \\ -2 \end{pmatrix}$$

5 Make a copy of the diagram below, and reflect the shape in the dotted line.

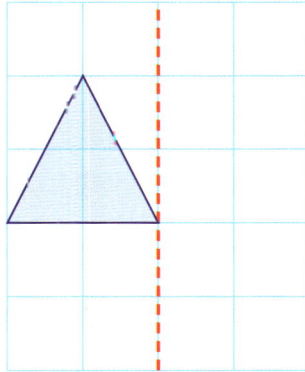

6 A point is at $(3, -3)$. What are its coordinates after reflection in the x-axis?

7 $ABCD$ is a square. The diagonals AC and BD meet at X. Name the triangles congruent to ABX.

Exercise 11B

1 Make a copy of the diagram below, and translate shape A 4 units to the right and 2 units up.

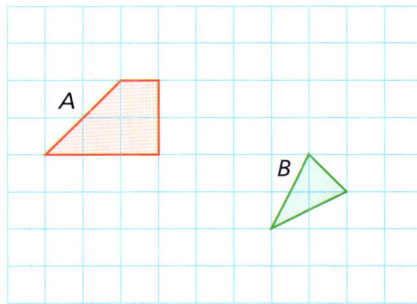

2 On your copy of the diagram for question 1, translate shape B 4 units to the left and 2 units down.

3 In the diagram below, what translation has taken triangle A to triangle B?

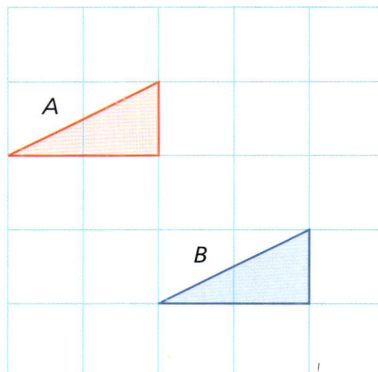

4 Write down the vector corresponding to the translation of question 2.

5 Make a copy of the diagram below, and reflect shape *A* in the horizontal dotted line.

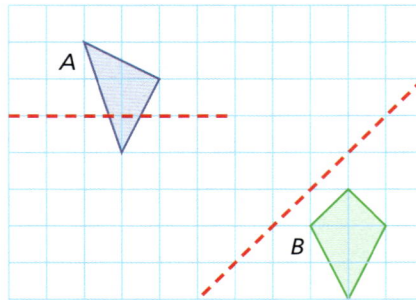

6 On your copy of the diagram for question 5, reflect shape *B* in the slanting dotted line.

7 A point is at $(3, 4)$. What are its coordinates after reflection in the line $y = x$?

8 A point is at $(4, 3)$. What are its coordinates after reflection in the line $y = 1$?

9 Which of the pairs of shapes below are congruent to each other?

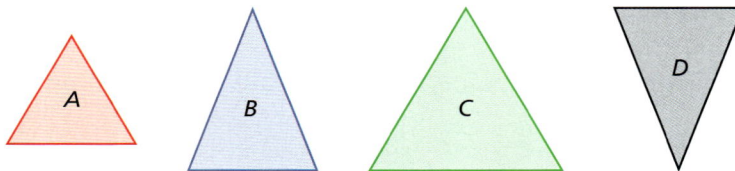

10 *ABCD* is a rectangle. The diagonals *AC* and *BD* meet at *X*. Name a triangle congruent to *AXB*.

Exercise 11C   Ma1

Have you ever been in a room which has mirrors on opposite walls? You see not one, but limitless numbers of reflections of yourself! There is the reflection of your reflection, the reflection of the reflection of your reflection, and so on endlessly.

What is the reflection of a reflection? Copy the diagram shown below. Reflect shape *A* in the line $x = 3$. Then reflect the result in the line $x = 0$. What is the result of the two reflections? Can you find a general rule? What happens when you go on reflecting?

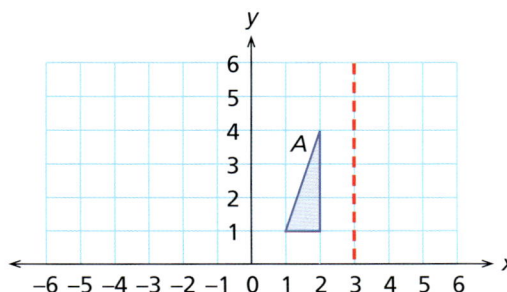

Exercise 11D   Ma1

A kaleidoscope is a tube that shows beautiful patterns when you look through it. At the end of the tube are several differently coloured shapes. These shapes are then reflected in mirrors, so that they seem to form a regular pattern.

The diagram below shows what the end of the kaleidoscope might be like. The three mirrors are at 60° to each other. Make a copy of the diagram, and reflect the shapes in the mirrors. Is the result a regular pattern?

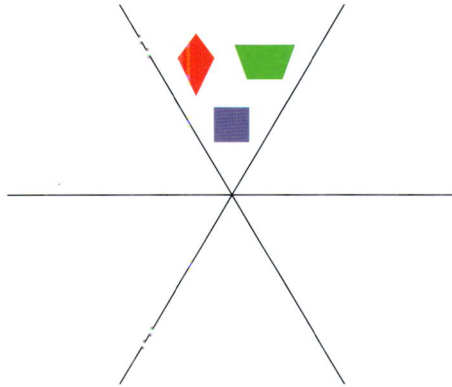

12 Sequences

Have you ever built a 'house of cards'? One type looks like a Chinese pagoda, with several storeys stacked on top of each other. How many cards are needed for each extra storey? How many storeys can you make with a pack of 52 cards?

Another type looks like a triangle, with fewer cards as you go higher. How many cards do you need for each storey now? How many storeys can you build with a pack of 52 cards?

You have already studied sequences of numbers. A sequence of numbers is a list of numbers which follow some pattern or rule. The sequence below starts with 2, and keeps on adding 3.

1st term	2nd term	3rd term	4th term	5th term	
2	5	8	11	14	...

The three dots at the end mean that the sequence goes on beyond 14. It can be continued for ever.

Example   What is the rule for this sequence?

1st term	2nd term	3rd term	4th term	5th term	
5	7	9	11	13	...

Each term is 2 greater than the one before. You could describe the sequence as

start at 5, then keep on adding 2

Exercise 12.1

(Mainly revision)
Find the next two terms for each of the sequences in questions 1 to 5.

1 2, 4, 6, 8, ...
2 10, 15, 20, 25, 30, ...
3 2, 5, 8, 11, 14, ...
4 22, 29, 36, 43, 50, ...
5 2, $2\frac{1}{2}$, 3, $3\frac{1}{2}$, 4, ...

6 to **10** Describe the rules for the sequences in questions 1 to 5.

Decreasing sequences

The sequences so far were all increasing. To go from one term to the next we added a positive number. A sequence could just as well be decreasing, so that to go from one term to the next we *subtract* a number.

Consider this sequence.

1st term	2nd term	3rd term	4th term	5th term	
25	21	17	13	9	...

Each term is 4 less than the one before. The next term is $9 - 4$, i.e. 5. The rule for this sequence could be

start with 25, keep on subtracting 4

Example   Find the next two terms of this sequence. Describe the rule for the sequence.

27, 22, 17, 12, 7, ...

Each term is 5 less than the one before. The fifth term is 7, so to find the sixth term subtract 5 from 7, obtaining 2. For the seventh term, subtract 5 from 2, obtaining -3.
The next two terms are 2 and -3.
The rule is

start at 27, keep on subtracting 5

Exercise 12.2

Find the next two terms for each of the sequences in questions 1 to 5.

1 17, 16, 15, 14, ...
2 21, 19, 17, 15, 13, ...
3 34, 27, 20, 13, 6, ...
4 2, $1\frac{1}{2}$, 1, $\frac{1}{2}$, 0, ...
5 2.3, 2.1, 1.9, 1.7, 1.5, ...

6 to **10** Find the rules for the sequences of questions 1 to 5.

Exercise 12.3 Ma1

Below are the numbers 1 to 100, written in a square. What sequences can you find in it? What if you go horizontally, or vertically, or diagonally? What if you go upwards or backwards?

1	2	3	4	5	6	7	8	9	10
11	12	13	14	15	16	17	18	19	20
21	22	23	24	25	26	27	28	29	30
31	32	33	34	35	36	37	38	39	40
41	42	43	44	45	46	47	48	49	50
51	52	53	54	55	56	57	58	59	60
61	62	63	64	65	66	67	68	69	70
71	72	73	74	75	76	77	78	79	80
81	82	83	84	85	86	87	88	89	90
91	92	93	94	95	96	97	98	99	100

Often you can find these sorts of sequences in day to day life!

Examples To hire a car, you pay £20 deposit, plus £30 per day. How much do you have to pay if you hire it for 1 day, for 2 days, for 3 days, and so on? Continue this sequence for two more days!

To hire for one day, you pay £20 + 1 × £30 = £50
To hire for two days, you pay £20 + 2 × £30 = £80
To hire for three days, you pay £20 + 3 × £30 = £110

This sequence is going up by £30 each time! For the next two days you only need to add £30.

To hire for four days, you pay £110 + £30 = £140
To hire for five days, you pay £140 + £30 = £170

At the beginning of the year I borrow £30, and pay it off at £3 every month. Write down how much I owe at the end of each of the first three months. Continue the sequence for two more months.

At the end of January, I owe $£30 - 1 \times £3 = £27$
At the end of February, I owe $£30 - 2 \times £3 = £24$
At the end of March, I owe $£30 - 3 \times £3 = £21$

This sequence is going down by £3 every month. Write down the next two lines.

At the end of April, I owe $£21 - £3 = £18$
At the end of May, I owe $£18 - £3 = £15$

Exercise 12.4

● ● ● ● ● ● ● ● ●

1 At the end of the year, Jonas's grandfather puts £40 in a savings account for him. At the end of every month Jonas is able to add £4. Find out how much is in the account at the end of each of the first three months. Continue the sequence for two more months.

2 Su Ying starts the year with £38 in her savings account, but every week draws out £5. Find how much she has left at the end of each of the first three weeks. Continue this sequence for two more weeks. Can the sequence go on for ever?

3 Rhoda is booking a holiday in France. The brochure quotes a price (including ferry crossing) of £203 for three nights. Each extra night costs £44. Copy and complete the table below.

number of nights	3	4	5	6	7
cost (£)	203	247			

4 Sonya is a keen gardener. She buys a plant which is 25 cm high. It is expected to grow by 8 cm each year for the next 7 years. Copy and complete the table below.

year	1	2	3	4	5	6	7
height (cm)	25	33					

Is it sensible to continue the table indefinitely?

5 In chapter 4, you found the sum of the angles of various polygons.

triangle	quadrilateral	pentagon	hexagon	heptagon	octagon
180°	360°				

Copy and complete the sequence. What is the sum of the angles of a nonagon (9 sides)?

Multiplying terms

Look at this sequence.

1st term	2nd term	3rd term	4th term	5th term
3	6	12	24	48 ...

It starts with 3. The second term is 3 greater. The third term is 6 greater than that. So you aren't adding the same amount on each time. Look carefully at the numbers:

$$6 = 2 \times 3 \quad 12 = 2 \times 6 \quad 4 = 2 \times 12 \quad 48 = 2 \times 24$$

Each term is double the one before. So each term is found by *multiplying* the one before by 2. The next term is 2×48, that is, 96. The rule could be

start with 3, keep on multiplying by 2

Examples Find the next two terms of this sequence. Describe the rule for the sequence.

$$5 \quad 15 \quad 45 \quad 135 \quad 405$$

Each term is three times the one before. Multiply 405 by 3, obtaining 1215. Multiply 1215 by 3, obtaining 3645.
  The next terms are 1215 and 3645.
  The rule is

start with 5, keep on multiplying by 3

I put £1000 into a bank account, at 5% interest per year. How much do I have at the end of each of the first three years?

At the end of the first year I have	£1000 × 1.05	£1050
At the end of the second year I have	£1050 × 1.05	£1102.50
At the end of the third year I have	£1102.50 × 1.05	£1157.625

Remember:
Increasing by 5% can be done by a multiplying factor of 1.05 (chapter 3).

Exercise 12.5

Find the next two terms of each of the sequences in questions 1 to 4.

1 2, 6, 18, 54, . . .
2 5, 10, 20, 40, . . .
3 2, 10, 50, 250, . . .
4 64, 96, 144, 216, . . .
5 to **8** Find the rules for the sequences in questions 1 to 4.

9 I invest £100 at 8% interest. Write down the amounts I have at the end of each of the first three years.

10 A colony of bacteria is growing on a plate. At 12 o'clock there are 2000 bacteria, and the number doubles every hour. Write down the numbers of bacteria at the end of each of the first four hours.

Dividing terms

Look at this sequence.

$$1 \quad \frac{1}{2} \quad \frac{1}{4} \quad \frac{1}{8} \quad \frac{1}{16} \quad \frac{1}{32} \quad \ldots$$

This sequence is getting smaller. But you are not subtracting the same amount each time. In fact, if you look carefully at these numbers, you can see that each term is found by *dividing* the one before by 2. The next term is $\frac{1}{32} \div 2$, that is $\frac{1}{64}$. The rule could be

start with 1, keep on dividing by 2

Example   Fishermen are reducing the stock of fish in a lake by 10% each year. Initially there are 100 000 fish. Find the number of fish at the end of each of the first three years.

If 10% of the fish are taken, then 90% remain. So the multiplying factor corresponding to a 10% decrease is 0.9.

At the end of the first year
the number of fish is $100\,000 \times 0.9$ 90 000
At the end of the second year
the number of fish is $90\,000 \times 0.9$ 81 000
At the end of the third year
the number of fish is $81\,000 \times 0.9$ 72 900

Exercise 12.6

Find the next two terms of each of the sequences in questions 1 to 4.

1 96, 48, 24, 12, ...
2 1458, 486, 162, 54, ...
3 $1, \frac{1}{3}, \frac{1}{9}, \frac{1}{27}, \ldots$
4 80, 20, 5, $1\frac{1}{4}$, ...
5 to **8** Find the rules for the sequences in questions 1 to 4.
9 A car is bought for £10 000, and at the end of each year it has lost 10% of its value at the start of the year. (This is known as depreciation.) Write down how much it is worth at the end of each of the next three years.
10 In a tropical country, over-foresting is causing the rain forest to disappear at 20% per year. Initially there were 120 000 square kilometres of rain forest. Find the amount of rain forest left at the end of each of the next three years.

Shape sequences

Look at the sequence of patterns below.

You can draw the next pattern, just by extending the arms of the L shape by one dot each.

Count the number of dots in each pattern. Write down your answer.

You should get the sequence below.

1   3   5   7   9

To continue the sequence, you don't need to draw more patterns. There must be 11 dots in the next pattern, 13 dots in the pattern after that, and so on. The number sequence tells you how the shape pattern increases.

Example   Look at the sequence of patterns below, made from sticks.

Draw the next pattern. Write down the numbers of sticks in each pattern, and continue this number pattern for two more terms.

The next pattern will have an extra square 'hole' in it.

The numbers of sticks in the pattern are

1   4   7   10   13

This sequence is going up in steps of 3. The next two terms will be 13 + 3, that is 16, then 16 + 3, that is, 19.
 The next two terms in the number sequence are 16 and 19.

Exercise 12.7

For each of questions 1 to 5, draw the next pattern, write down the numbers in each of the patterns, and continue the number sequence for two more terms.

1

2

3

4

5

Method of differences

Look at this sequence.

$$1 \quad 3 \quad 6 \quad 10 \quad 15 \quad 21 \quad \dots$$

This sequence is increasing. The second term is 2 greater than the first, and the third term is 3 greater than the second. So the sequence isn't found by adding the same amount each time.

What about multiplication? The second term is 3 times the first term, and the third term is 2 times the second term. So the sequence isn't found by multiplying by the same amount each time.

With a sequence like this, it often helps to write down the *differences* of the terms. These are given by the table below.

term	1	3	6	10	15	21
difference	2	3	4	5	6	

Now there is a pattern. The differences are increasing by 1 each time. The next term along will be 7 greater than 21, which is 28. The term after that will be 8 greater than 28, which is 36.

Example   Find the next two terms of the sequence 2, 5, 10, 17, 26, . . .

Draw up a table of differences, as below.

term	2	5	10	17	26
difference	3	5	7	9	

The differences are 3, 5, 7, 9. These are going up by 2 each time. So the next difference is 11. Add this to 26, obtaining 37. The next difference is 13. Add this to 37, obtaining 50.

The next two terms are 37 and 50.

Exercise 12.8

Find the next two terms for each of these sequences.

1 2, 3, 5, 8, 12, . . .
2 0, 2, 6, 12, 20, . . .
3 1, 3, 7, 13, 21, . . .
4 4, 7, 11, 16, 22, . . .
5 1, 1.1, 1.3, 1.6, 2.0, . . .

6 1, $1\frac{1}{2}$, $2\frac{1}{2}$, 4, 6, . . .
7 1, 1, 2, 4, 7, . . .
8 5, 3, 2, 2, 3, . . .
9 25, 24, 22, 19, 15, . . .
10 20, 16, 13, 11, 10, . . .

Exercise 12.9 Ma1

Below are the numbers 1 to 21, written in a triangle. What sequences can you find in the triangle?

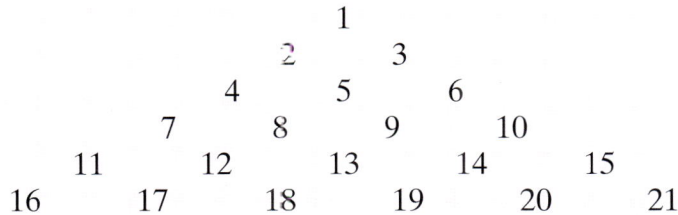

```
                    1
                2       3
            4       5       6
        7       8       9       10
    11      12      13      14      15
16      17      18      19      20      21
```

Example The diagram below shows a sequence of triangles made from dots. Draw the next pattern. Write down the corresponding number sequence, and continue it for two more terms.

The next pattern has an extra row of dots, as shown.

The numbers of dots in the patterns are

1   3   6   10   15

Remember:

These are called the **triangle numbers**.

The differences between these numbers are 2, 3, 4 and 5. The next difference is 6, so the next term is 21. The term after that is 21 + 7, that is, 28.

The next two terms are 21 and 28.

Exercise 12.10

For each of questions 1 to 5, find the next pattern, write down the numbers in each of the patterns, and continue the number sequence for two more terms.

1

2

 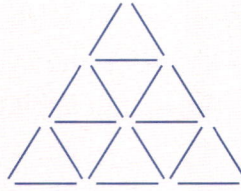

3

4

5

6 The diagram below shows a sequence of regular polygons. A **diagonal** of a polygon is a line joining two vertices which aren't next to each other. Copy and complete the table below.

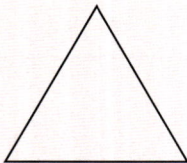  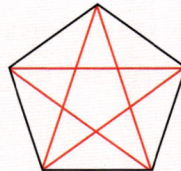 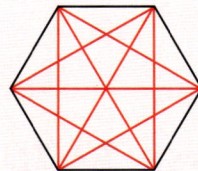

number of sides	3	4	5	6	7	8	9
number of diagonals	0	2	5				

7 The diagram below shows circles with points spaced evenly round the circumference. For each circle, all the points are joined up to each other. How many lines are there? Copy and complete the table below.

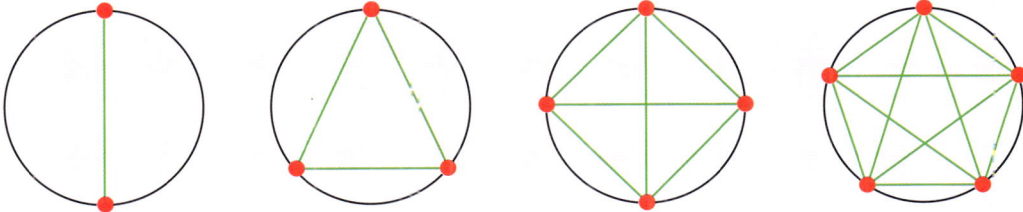

number of points	2	3	4	5	6
number of lines		3	6		

8 Look again at question 7. Think of the circles as pizzas, and of the lines as cuts. How many pieces of pizza do you have in each case? Copy and complete the table below.

number of points	2	3	4	5	6
number of pieces	2	4			

Exercise 12.11

At the beginning of the chapter we had two types of 'house of cards', one like a pagoda, one with a triangle shape. For each type, copy and complete the table below.

	pagoda	triangle
number of cards for 1 storey		
number of cards for 2 storeys		
number of cards for 3 storeys		

Can you spot the patterns? Continue the sequences without actually counting the numbers of cards.

Finding the rule

We have had several sorts of sequences: in some you added a fixed number, in some you subtracted, in some you multiplied or divided. In some sequences the *differences* between terms were increasing or decreasing in a regular way. If you are given a sequence, here's how you could investigate to find out what sort of rule it is following.

- Is the sequence increasing or decreasing?
- Suppose it is increasing. Is the amount by which it is increasing the same each time?
- Find the ratio between successive terms. Is this always the same?
- Write out the differences between successive terms. Are they increasing or decreasing by a constant amount?

After following through these steps you will have found out what sort of sequence it is, and you will be able to continue it for the next few terms.

Exercise 12.12

Find the rules for each of these sequences and then write down the next two terms.

1 4, 7, 10, 13, . . .
2 7, 14, 28, 56, . . .
3 9, 7, 5, 3, . . .
4 10, 15, 25, 40, . . .
5 162, 54, 18, 6, . . .

6 4, 4, 6, 10, 16, . . .
7 5.6, 5.1, 4.6, 4.1, . . .
8 $\frac{1}{6}$, $\frac{1}{2}$, $\frac{5}{6}$, $1\frac{1}{6}$, $1\frac{1}{2}$, . . .
9 0.5, 1.5, 4.5, 13.5, . . .
10 243, 162, 108, 72, . . .

The general term

A sequence has a first term, a second term, a third term, and so on. If we want the hundredth term, it is a nuisance to have to find all the 99 terms before it! Sometimes it is possible to find an expression for the **general term**, that is, the **nth term**. This is an expression in n.

1st term   2nd term   3rd term   . . . nth term   . . .

To find the hundredth term, we just substitute 100 for n.

Examples   The diagram below shows a sequence of patterns of dots. Find an expression for the number of dots in the nth pattern. How many dots are there in the 20th pattern?

There are 4 dots in the first pattern, 8 dots in the second, 12 dots in the third, and so on.

	1st pattern	2nd pattern	3rd pattern	4th pattern
number of dots	4	8	12	16
	4 × 1	4 × 2	4 × 3	4 × 4

At each stage we multiply by 4. So for the nth pattern, there are $4 \times n$ dots.

The nth pattern has $4n$ dots.
For the 20th pattern, there are 4×20 dots.
The 20th pattern has 80 dots.

The diagram below shows a sequence of patterns of squares. Find the number of squares in the nth pattern.

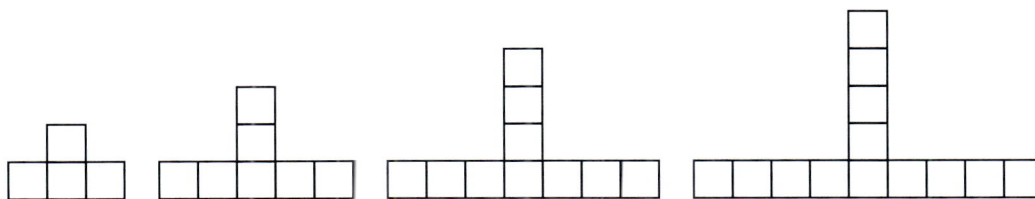

There are 4 squares in the first pattern, 7 in the second, 10 in the third, and so on.

	1st pattern	2nd pattern	3rd pattern	4th pattern
number of squares	4	7	10	13

The number of squares is going up by 3 each time. But notice that there is always an extra square in the middle. The number of squares can be written as

	1st pattern	2nd pattern	3rd pattern	4th pattern
number of squares	3 + 1	3 × 2 + 1	3 × 3 + 1	3 × 4 + 1

So by the nth term, the expression is $3 \times n + 1$
The nth term is $3n + 1$.

Exercise 12.13

For questions 1 to 3, find an expression for the number of dots in the *n*th pattern.
How many dots are there in the 20th pattern?

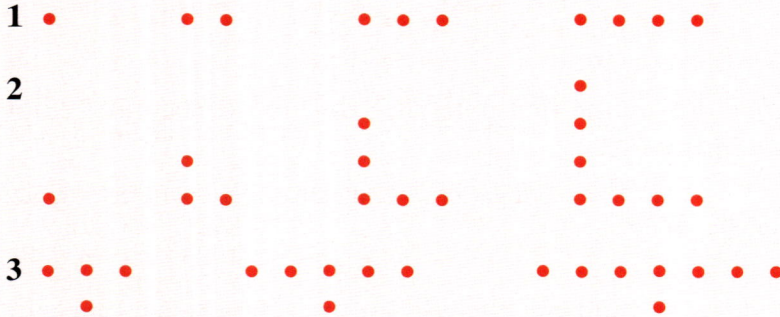

1

2

3

For questions 4 to 6, find an expression for the number of lines in the *n*th pattern.
How many lines are there in the 30th pattern?

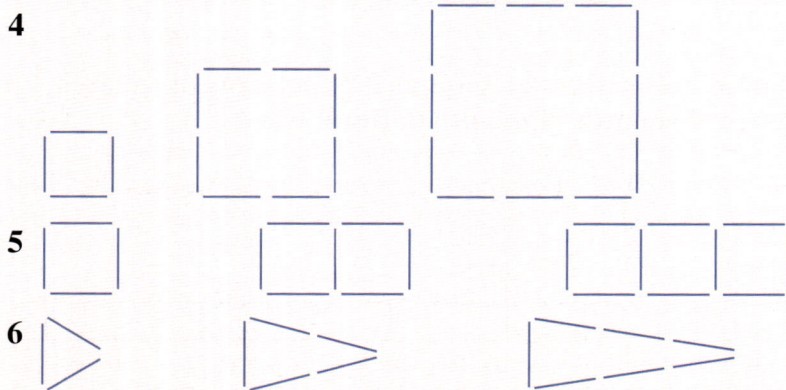

4

5

6

For questions 7 and 8, find an expression for the number of squares in the *n*th pattern.

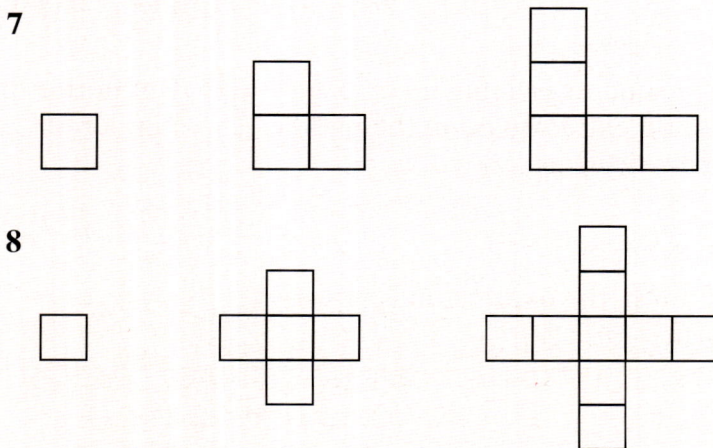

7

8

9 Mavis is making a sequence of patterns out of matchsticks, as shown below. How many matchsticks are needed for a pattern with *n* triangles?

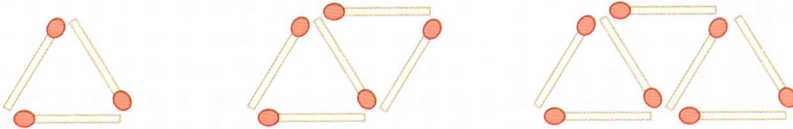

10 A fence is made with regularly spaced vertical pieces of wood, joined by horizontal pieces of wood, as shown below. How many pieces of wood are needed for a fence with *n* verticals?

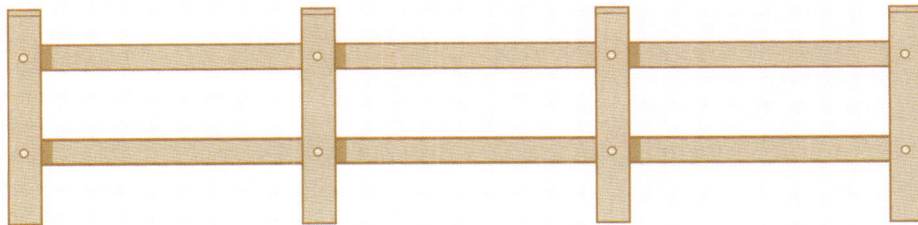

SUMMARY

- A number sequence is a list of numbers which follow a pattern or rule. The rule tells us how to go from one term to the next.
- The rule of a sequence might be to add or subtract a fixed number to the previous term, or to multiply or divide the previous term by a fixed number.
- A sequence of shapes may have an associated number sequence.
- In some sequences the *differences* between the terms increase or decrease at a fixed rate.
- For some sequences you can find an expression for the *n*th term.

Exercise 12A

1 Continue the sequence 3, 7, 11, 15, . . . for two more terms.
2 Describe the rule for the sequence in question 1.
3 Continue the sequence 101, 93, 85, 77, . . . for two more terms.
4 Describe the rule for the sequence in question 3.
5 Continue the sequence 1024, 768. 576, 432, . . . for two more terms.
6 Describe the rule for the sequence in question 5.
7 Continue the sequence 0, 3, 8, 15, . . . for two more terms.
8 A new town starts with a population of 100 000, and it increases at 20% each year. Find the population at the end of each of the first three years.
9 I start the week with 23 bottles of mineral water, and drink three bottles per day. How many bottles are there left at the end of each of the first three days? Continue the sequence until the end of the week.

10 Draw the next pattern in the sequence below. Write down the corresponding number sequence for the number of lines, and continue it for two more terms. Find an expression for the *n*th term.

Exercise 12B

1 Continue the sequence $2\frac{1}{2}$, $2\frac{3}{4}$, 3, $3\frac{1}{4}$, $3\frac{1}{2}$, . . . for two more terms.

2 Describe the rule for the sequence in question 1.

3 Continue the sequence 1, 10, 100, 1000, 10 000, . . . for two more terms.

4 Describe the rule for the sequence in question 3.

5 Continue the sequence 1, 0.2, 0.04, 0.008, 0.0016, . . . for two more terms.

6 Describe the rule for the sequence in question 5.

7 Continue the sequence 10, 5, 5, 10, 20, 35, . . . for two more terms.

8 A rubbish tip is 2 m high. Every week an extra 0.1 m is added to it. Find the height of the tip at the end of each of the next five weeks.

9 A government hopes that the number of unemployed people will decrease by 10% each year. At the moment there are 1 200 000 unemployed. Write down the expected number of unemployed at the end of each of the next four years.

10 Draw the next pattern in the sequence on the right. Write down the corresponding number sequence for the number of dots, and continue it for two more terms.

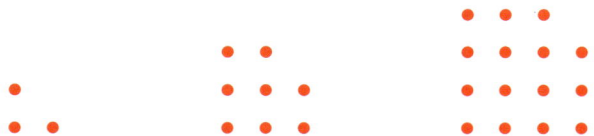

Exercise 12C   Ma1

On the right is part of the addition table. Copy, complete it and extend it for two more rows. You don't have to do any more addition! You can do it by counting!

+	1	2	3	4	5	6	7	8	9
1	2	3	4	5					
2	3	4	5						
3	4	5	6						

On the right is part of the multiplication table. Copy, complete it and extend it for two more rows. You don't have to do any more multiplication! You can do it by addition!

+	1	2	3	4	5	6	7	8	9
1	1	2	3	4					
2	2	4	6						
3	3	6	9						

Exercise 12D Ma1

Take several sheets of paper; draw one line across the first sheet, two lines across the second, three across the third, and so on, as shown below. Make sure that all the lines cross each other, and that you don't have more than two lines going through the same point.

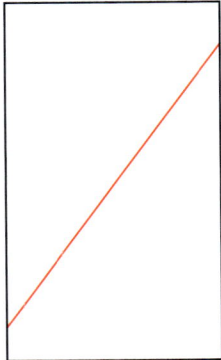

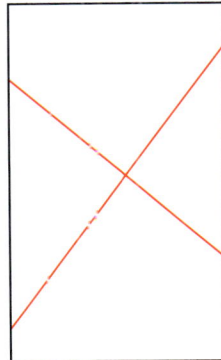

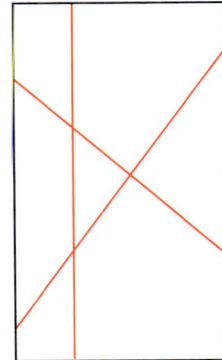

1   How many crossing points are there?
2   Find the sequence of numbers of crossing points.
3   How many separate regions of the paper are there?
4   Find the sequence of numbers of regions.

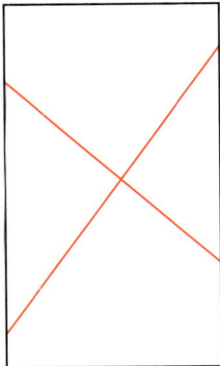

1 crossing point

4 regions

13 : Powers and indices

Have you or anyone in your family bought a computer? Do you know how much memory it has? The basic unit of memory is the **byte**, which can hold one letter or one digit. Obviously you need more than one byte; the next unit of memory is the **kilobyte**.

$$1 \text{ kilobyte} = 1024 \text{ bytes}$$

The amount of memory in computers has increased rapidly over the past few years. In the early 1980s the BBC Model A computer had a memory of 16 kilobytes, and the Model B a memory of 32 kilobytes. The Amstrad PCW had a memory of 256 kilobytes.

To run the Windows 98 operating system, a computer needs a memory of at least 16 megabytes.

$$1 \text{ megabyte} = 1024 \text{ kilobytes}$$

For the hard disk of a personal computer, you need even more memory. The size of this memory is often given in **gigabytes**.

$$1 \text{ gigabyte} = 1024 \text{ megabytes}$$

And so it goes on. It can't be long before memory is measured in terabytes.

$$1 \text{ terabyte} = 1024 \text{ gigabytes}$$

Look again at the numbers above: 16, 32, 256 and 1024. Do you notice anything in common?

Indices

You already know a way of writing a number when it is multiplied by itself.

$$3 \times 3 = 3^2$$

This is '3 squared', or '3 to the power 2'. It gives the area of a square of side 3 units. The number 2 is the **index**. The number 3 is the **base number**.

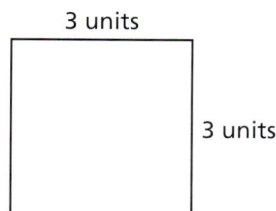

Remember:

base ——— 3^2 ——— index
number

3 units

3 units

Similarly we write $4 \times 4 \times 4$ as 4^3. This is '4 cubed', or '4 to the power 3'. It gives the volume of a cube of side 4 units. Here the index is 3 and the base number is 4.

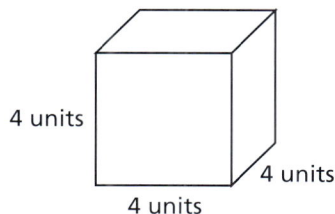

4 units
4 units
4 units

Exercise 13.1

(Mainly revision)
In questions 1 to 5, evaluate the expressions.

1 2^2
2 5^2
3 2^3
4 3^3
5 3^4

In questions 6 to 9, write the expressions using indices.

6 7×7
7 10×10
8 $2 \times 2 \times 2$
9 $3 \times 3 \times 3 \times 3$
10 When you read the introduction to this chapter, did you find anything in common with the numbers involved: 16, 32, 256 and 1024? In fact, they are all powers of 2.

$$16 = 2^4 \qquad 256 = 2^8$$

What are 32 and 1024 as powers of 2?

Using base letters instead of numbers

In Book 1 and in the exercise above you found powers of numbers. We can be more general if we use letters to stand for the numbers.

$$4 \times 4 \times 4 = 4^3 \qquad x \times x \times x = x^3$$

Just write letters where previously you wrote numbers.

Exercise 13.2

Write the following as powers of a single letter.

1 $c \times c \times c \times c \times c$

2 $y \times y \times y$

3 $x \times x \times x \times x$

4 $u \times u$

5 $a \times a \times a \times a \times a$

6 $t \times t \times t$

7 $n \times n \times n \times n \times n \times n \times n$

8 $x \times x$

9 $f \times f \times f \times f$

10 $p \times p \times p \times p \times p \times p \times p \times p$

Exercise 13.3

Write out these powers in full as products.

1 x^6

2 a^3

3 q^4

4 s^9

5 w^2

6 m^5

7 r^8

8 b^2

9 z^7

10 x^1

The last one looks like a trick question! The power x^1 is equal to x, as it isn't multiplied by any other x!

Multiplication of powers

Take some simple powers.

$$a^2 = a \times a \qquad a^3 = a \times a \times a$$

Now multiply them together.

$$a^2 \times a^3 = \underbrace{a \times a}_{a^2} \times \underbrace{a \times a \times a}_{a^3} = a^5$$

So we multiplied powers of a, and the result was another power of a. Now you try some.

Exercise 13.4

1 a Write out k^4 in full.

  b Write out k^2 in full.

  c Use parts **a** and **b** to write out $k^4 \times k^2$ in full

  d Hence write $k^4 \times k^2$ as a power of k.

2 a Write out d^5 in full.
  b Write out d^3 in full.
  c Use parts **a** and **b** to write out $d^5 \times d^3$ in full.
  d Hence write $d^5 \times d^3$ as a power of d.

3 a Write out g^6 in full.
  b Write out g^2 in full.
  c Use parts **a** and **b** to write out $g^6 \times g^2$ in full.
  d Hence write $g^6 \times g^2$ as a power of g.

You should be able to see a pattern in the indices.

$$b^2 \times b^3 = \underbrace{b \times b}_{2\ bs} \times \underbrace{b \times b \times b}_{3\ bs} = \underset{5\ bs}{b^5}$$

$$t^4 \times t^5 = \underbrace{t \times t \times t \times t}_{4\ ts} \times \underbrace{t \times t \times t \times t \times t}_{5\ ts} = \underset{9\ ts}{t^9}$$

Invent an example like this and write it in your book. Check it with a partner.

Exercise 13.5

Write the answers to the following products in the simplest way you can.

1 $x^3 \times x^2$	**3** $z^3 \times z^3$	**5** $b^2 \times b^4$
2 $y^4 \times y^2$	**4** $a^4 \times a^1$	**6** $k^4 \times k$

Look carefully at the questions and answers to exercise 13.5 and write a sentence explaining the rule when you multiply powers.

Exercise 13.6

Use your rule to write down the answers to these products, in index form.

1 $a^3 \times a^5$	**3** $r^4 \times r^2$	**5** $u^3 \times u^5$
2 $b^2 \times b^3$	**4** $n^5 \times n^3$	**6** $v^8 \times v$

Putting together all the previous work, you should have noticed this rule.

● *To multiply powers, you add their indices.*

Exercise 13.7   Ma1

At the beginning of this chapter we mentioned bytes, kilobytes, megabytes and terabytes.

1 terabyte = 1024 gigabytes
1 gigabyte = 1024 megabytes
1 megabyte = 1024 kilobytes
1 kilobyte = 1024 bytes

In question 10 of exercise 13.1 you should have found that 1024 is 2^{10}.
In these questions give your answers as powers of 2.

1 How many bytes are there in a megabyte?
2 How many bytes are there in a gigabyte?
3 How many bytes are there in a terabyte?

Dividing powers

Look carefully at what happens when 6^5 is divided by 6^2.

$$6^5 \div 6^2 = \frac{6 \times 6 \times 6 \times 6 \times 6}{6 \times 6}$$

Divide numerator and denominator by any common factors.

$$6^5 \div 6^2 = \frac{6 \times 6 \times 6 \times \cancel{6} \times \cancel{6}}{\cancel{6} \times \cancel{6}}$$

$$= \frac{6 \times 6 \times 6}{1}$$

$$= 6^3$$

Here's another example, using letters.

$$y^7 \div y^3 = \frac{y \times y \times y \times y \times y \times y \times y}{y \times y \times y}$$

$$y^7 \div y^3 = \frac{y \times y \times y \times y \times \cancel{y} \times \cancel{y} \times \cancel{y}}{\cancel{y} \times \cancel{y} \times \cancel{y}}$$

$$= \frac{y \times y \times y \times y}{1}$$

$$= y^4$$

Exercise 13.8

Work out the answers to these divisions of powers. Do the first five like the examples above, but after that try to shorten the working.

1 $3^5 \div 3^2$

2 $4^6 \div 4^3$

3 $x^7 \div x^4$

4 $y^5 \div y^2$

5 $z^8 \div z^5$

6 $6^5 \div 6^2$

7 $8^7 \div 8^3$

8 $m^6 \div m^2$

9 $p^8 \div p^3$

10 $n^7 \div n$

Have you spotted the rule?

Write a sentence about division of powers, explaining the rule when you divide two powers.

Exercise 13.9

Use your result to simplify these expressions as quickly as possible.

1 $7^6 \div 7^3$

2 $8^9 \div 8^1$

3 $9^5 \div 9^2$

4 $9^{12} \div 9^5$

5 $3^{18} \div 3^{13}$

6 $x^8 \div x^5$

7 $y^9 \div y^4$

8 $p^{15} \div p^4$

9 $q^{23} \div q^{13}$

10 $r^{31} \div r$

The rule you should have found is this.

● *When dividing powers, you subtract the indices.*

Exercise 13.10

In exercise 13.7 you should have found the following connection between bytes, kilobytes, megabytes, gigabytes and terabytes.

1 kilobyte = 2^{10} bytes   1 megabyte = 2^{20} bytes
1 gigabyte = 2^{30} bytes   1 terabyte = 2^{40} bytes

1 How many kilobytes are there in a megabyte?

2 How many kilobytes are there in a gigabyte?

3 How many kilobytes are there in a terabyte?

4 How many megabytes are there in a gigabyte?

5 How many megabytes are there in a terabyte?

6 How many gigabytes are there in a terabyte?

Multiplying with different base numbers and letters

Remember:

In the expression 3^4, the index is 4 and the base number is 3.

So far in this chapter, each multiplication or division of powers has involved only *one* base number or letter. Look back over the previous questions to check this.

This is important. Expressions like $2^3 \times 2^2$ can be simplified to 2^5, but expressions like $2^3 \times 3^2$ cannot be written as a single power, because the base numbers are different.

So an expression like $a^3 \times b^2$ cannot be simplified any further, for the same reason.

Exercise 13.11

Simplify the following *where possible*.

It may not be possible to simplify!

1 $3^3 \times 3^4$	**4** $2^2 \times 2^3$	**7** $a^3 \times b^2$	**9** $b^5 \times c^4$
2 $2^3 \times 5^2$	**5** $4^3 \times 3^2$	**8** $b^5 \times b^3$	**10** $a^3 \times b^2 \times c^4$
3 $2^2 \times 3^3$	**6** $a^3 \times a^2$		

Matching base letters

As we have seen, when you multiply or divide powers, if the base numbers match, then you can write the result as a single power. The same applies to powers of letters. If the base letters match, some simplification can be done. If you have a multiplication involving several powers, you can simplify it if some of the base numbers match. You may have to rearrange the multiplication.

Examples Simplify $x^3 \times x^5 \times y^4$.

$$x^3 \times x^5 \times y^4 = x^8 \times y^4 \qquad \text{(combining the } x\text{s)}$$

Simplify $x^7 \times y^3 \times x^2$.

$$x^7 \times y^3 \times x^2 = x^7 \times x^2 \times y^3 \qquad \text{(bringing the } x\text{s together)}$$
$$= x^9 \times y^3 \qquad \text{(combining the } x\text{s)}$$

The matching letters are brought together.

Exercise 13.12

Simplify the following as far as possible.

1 $x^2 \times x^4 \times z^5$

2 $a^3 \times a^2 \times b^3$

3 $p^4 \times q^2 \times q^5$

4 $a^2 \times b^3 \times a^5$

5 $x^3 \times y^3 \times x^8$

6 $x^3 \times x^5 \times y^3 \times y^3$

7 $a^2 \times a^4 \times b^3 \times b^5$

8 $x^5 \times y^7 \times y^3 \times x^2$

9 $a^4 \times b^6 \times a^3 \times b^2$

10 $p^2 \times q^4 \times p^2 \times q^8$

Adding powers: collecting like terms

Remember chapter 5.

We can simplify additions in algebra when they contain *like terms*. If they contain unlike terms, they can't be simplified.

$$3a + 5a = 8a \qquad \text{(they are like terms)}$$
$$3a + 5b \text{ cannot be simplified} \qquad \text{(they are unlike terms)}$$

The same sort of thing holds when adding powers. You can add powers of x, provided they are the same power. You can *multiply* powers if they have the same base letter. But you cannot *add* powers unless they have the same base letter *and* the same power. Only then can they be considered 'like terms'.

$$3x^2 + 4x^2 = 7x^2 \qquad \text{(they are like terms)}$$
$$3x^2 + 4y^2 \text{ cannot be simplified} \qquad \text{(they are unlike terms)}$$
$$3x^2 - 4x^3 \text{ cannot be simplified} \qquad \text{(they are unlike terms)}$$

Exercise 13.13

Simplify these expressions as far as possible.

1 $2x^3 + 5x^3$

2 $3x^3 + 4x^4$

3 $5y^3 + 7y^3$

4 $3y^3 + 7x$

5 $7p^2 + 8p^2$

6 $q^3 + q^3$

7 $x^4 + 4x^4$

8 $6y^3 + y^2$

9 $7x^6 - 3x^6$

10 $5w^5 - 2w^5$

SUMMARY

- $6 \times 6 \times 6$ is written as 6^3. It is '6 to the **power** 3'. It has the value 216. The **base number** is 6 and the **index** is 3.
- To multiply powers you add the indices.
- To divide powers you subtract the indices.
- When you use the rules about multiplying or dividing powers, make sure that they have the same base number (or letter)
- When you add powers, make sure that they have the same base letter and the same power.

Exercise 13A

1 Work out the value of 5^4.
2 Write $7 \times 7 \times 7 \times 7 \times 7$ using indices.
3 Write f^3 in full.
4 Write $h \times h \times h \times h \times h \times h$ as a power of h.
5 Simplify $r^5 \times r^3$.

6 Simplify $r^5 \div r^4$.
7 Write 729 as a power of 9.
8 Simplify $5x^8 + 7x^8$.
9 Simplify $2x^3 + 3x^4$ as far as possible.
10 Simplify $a^4 \times b^8 \times a^2 \times b^9$.

Exercise 13B

1 Evaluate 10^4.
2 Write $12 \times 12 \times 12 \times 12 \times 12$ as a power of 12.
3 Write g^5 in full.
4 Put $j \times j$ in index form.
5 Write $k^2 \times k^6$ in a shorter way.

6 Simplify $t^{10} \div t^3$.
7 What is the square of 10^2?
8 A *googol* is 10^{100}. What is the square of a googol?
9 Simplify $a^2 \times c \times b^5 \times c^4 \times a^2 \times b$.
10 Simplify $3x^4 + 2x^4 + 4x^3$.

Exercise 13C   Ma1

A milligram (mg) is a thousandth of a gram. So there are 10^3 milligrams in 1 gram.

1 A microgram (μg) is a thousandth of a milligram, so it is a millionth of a gram. How many micrograms are there in a gram? Write your answer as a power of 10.
2 A nanogram (ng) is a thousandth of a microgram. How many nanograms are there in a gram? Write your answer as a power of 10.
3 A picogram (pg) is a thousandth of a nanogram. How many picograms are there in a gram? Write your answer as a power of 10.
4 A femtogram (fg) is a thousandth of a picogram. How many femtograms are there in a gram? Write your answer as a power of 10.

There is an even smaller unit! Can you find it?

Exercise 13D · Ma1

The number seven occurs over and over again in legends, proverbs, sayings and superstitions. Powers of seven also occur! This is a well-known nursery rhyme.

> As I was going to St Ives,
> I met a man with seven wives,
> Every wife had seven sacks,
> Every sack had seven cats,
> Every cat had seven kits
> Kits, cats, sacks, wives
> How many were going to St Ives?

1 How many kits were there? Give your answer as a power of seven.

This rhyme is very similar to a problem in an ancient Egyptian document, called the *Rhind Papyrus*, dating from about 1650 BC.

> There are seven houses,
> Each house has seven cats,
> Each cat ate seven mice,
> Each mouse would have eaten seven grains
> Each grain would have grown to seven *hekats* of corn.

2 The *hekat* was a measure of corn. How many *hekats* were saved? Give your answer as a power of seven.

There is a tradition of the *Sufis* (an Iranian mystic sect) about heavenly creatures, each with 70 000 heads, each head with 70 000 faces, each face with 70 000 mouths, each mouth with 70 000 tongues, each tongue speaking 70 000 languages.

3 How many languages could one of these creatures speak? Give your answer as a power of 70 000. Can you write your answer in terms of a power of 7 and a power of 10?

14 Straight-line graphs

They do things differently abroad! When you go on holiday outside Britain, there are many things you have to change. One is money – you have to adjust to the different currency. How do you do it?

If you are taking a car abroad, there are other conversions to do such as between miles and kilometres. Also if you want to check your tyre pressure, you may need a conversion between pounds per square inch and kilograms per square centimetre.

We have already looked at coordinates. The position of a point can be given by two numbers: its x-coordinate and its y-coordinate. This idea can be extended to show a connection between two quantities. The x-axis represents the first quantity, and the y-axis the second quantity. A point on the graph shows a connection between the x quantity and the y quantity. We can use this connection to help us do conversions.

Remember chapter 6.

Remember:

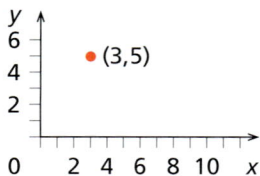

Conversion graphs

When you go abroad, you need to convert your British money into the local currency. There are several ways you could work out how much you are going to get.

You could use a table, showing how much you get for different sums of money.

number of pounds	number of Swiss francs
1	2.5
2	5
10	25

You could use a formula, such as

number of Swiss francs = 2.5 × number of pounds (£)

You could prepare a graph, so that the amount of money can be read off directly. This is a **conversion graph**, because it converts pounds sterling to Swiss francs. It might look like the diagram below. Along the x-axis are amounts in pounds (£), and up the y-axis are amounts in Swiss francs (SF). A point on the graph tells us that a certain amount in £ can be changed to a certain amount in SF.

If you change £18, how many SF will you get?

Start at 18 on the *x*-axis, go up to the graph, then go across to the *y*-axis. You see that you will get 45 SF.

At the end of the holiday you find you have 20 SF left. How much is that in pounds?

Start at 20 on the *y*-axis, go across to the graph, then down to the *x*-axis. You see that you get £8.

Sometimes reading the graph is more difficult. If you change £15, how many SF will you get? Reading from the graph, you get something between 37 SF and 38 SF, which you estimate as 37.5 SF. This is not completely accurate but is good enough for our purposes.

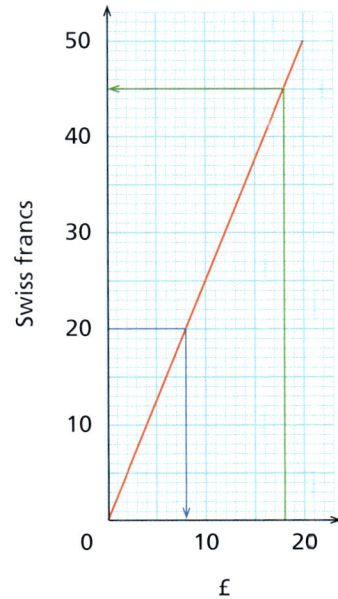

Exercise 14.1

1 The diagram below converts between pounds sterling (£) and US dollars ($). Use the graph to find the following.

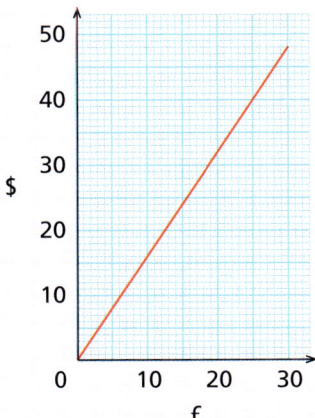

a The number of dollars you get for £15
b The number of pounds you get for $40

2 The diagram on the right is a conversion graph
for pounds weight (lb) and kilograms (kg).
Use the graph to find the following.
a The number of kilograms equivalent to 15 lb
b The number of pounds weight equivalent to 10 kg

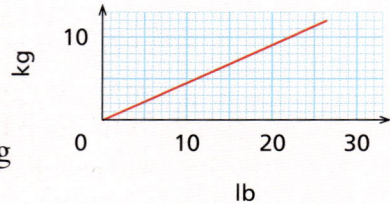

3 The graph on the right converts between litres
and gallons. Use the graph to find the following.
a The number of litres in a 5 gallon barrel
b The number of gallons in a 30 litre tank

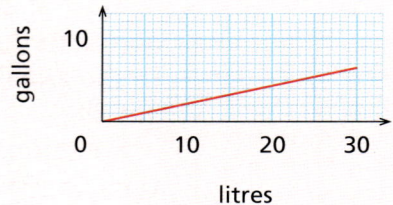

4 Many goods, such as computers, are advertised with prices *before* VAT has been
added. The graph below converts between the price before VAT and the price
including VAT.

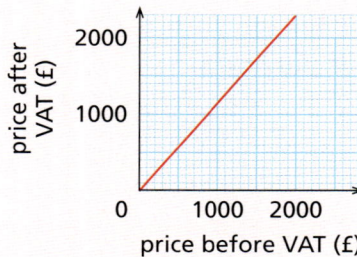

a Use the graph to find the price, including VAT, of a computer advertised as
£850 before VAT.
b I can afford £1500 for a computer. What price (before VAT) of computer can I
afford?

5 The fuel economy of a car can be measured in miles per gallon (m.p.g.) or in
kilometres per litre (km/litre). Below is a conversion graph for them.

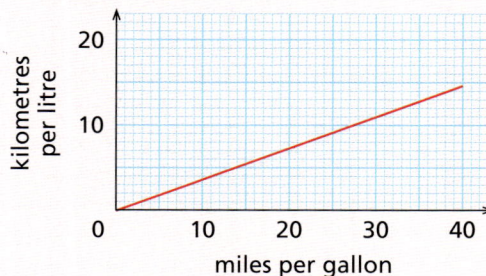

a A 30 year old 'Leviathan' car can do 15 miles per gallon. What is that in
kilometres per litre?
b A new 'Titania' car can do 14 km per litre. What is that in m.p.g.?

Making conversion graphs

Before you go on holiday, you might want to prepare a graph so that you can quickly find out how much things cost. This is how you could go about it.

- Set up a table. giving the foreign equivalent of various sums of British money.

pounds	French francs
1	10
2	20

- Draw axes on graph paper, with the £ along the *x*-axis and the foreign currency up the *y*-axis.

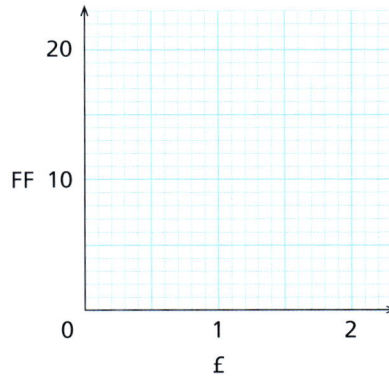

- Plot points on the graph, using the values found in the table. Join them up by a straight line.

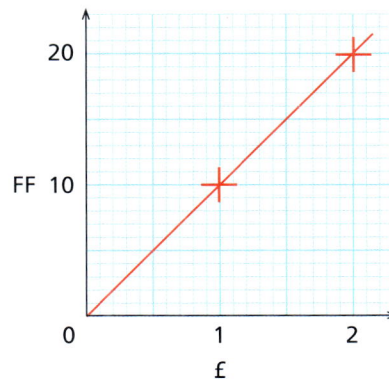

Example Suppose you are going to South Africa, where the exchange rate is 8 Rand to the pound. Find the South African equivalents of £0, £20 and £50. Prepare a conversion graph which you can use to convert between Rand and £.

For the conversion, multiply the number of £ by 8, to give the amount in Rand. The results are below.

amount (£)	0	20	50
amount (Rand)	0	160	400

To prepare a conversion graph, use graph paper, and mark the *x*-axis from 0 to 50, in divisions representing £10. Mark the *y*-axis from 0 to 400, each division representing 100 Rand. Plot the points on the graph, then join up by a straight line, as shown.

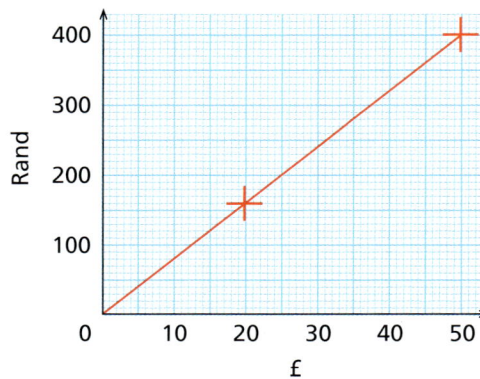

If the three points aren't in a straight line, then you have made a mistake. Either your arithmetic is wrong, or you plotted a point in the wrong place.

For a straight line, you only need two points. But it is worth having a third point, as a check.

Exercise 14.2

1 Patrick is going to Australia, where the exchange rate is 2.5 Australian dollars ($) to the British pound (£). Copy and fill in the table below. Make a copy of the diagram below on graph paper, and use it to prepare a conversion graph for £ and Australian $.

amount (£)	0	100	200
amount (Aus. $)	0		

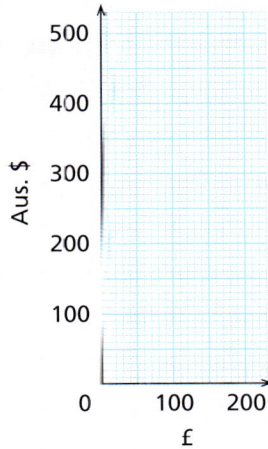

2 At a certain time there were 200 Japanese Yen (¥) to the £. Copy and complete the table below, and use it to prepare a conversion graph between pounds and Yen. Use a copy of the graph below.

amount (£)	0	100	200
amount (¥)	0		

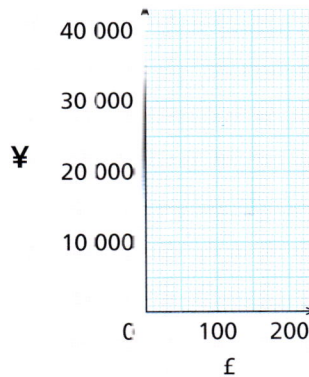

3 There are 0.4 hectares in one acre. Copy and complete the table below, and use it to prepare a conversion graph for hectares and acres. Use a copy of the graph below.

number of acres	0	10	20
number of hectares			

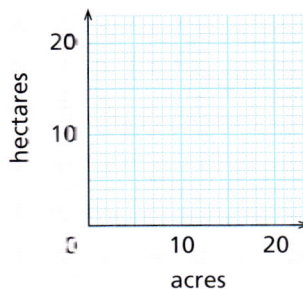

4 Suppose the rate of VAT becomes 25%. Copy and complete the table below to show prices before and after VAT. Prepare a conversion graph, using a copy of the graph below.

price before VAT (£)	0	100	200
price including VAT (£)		125	

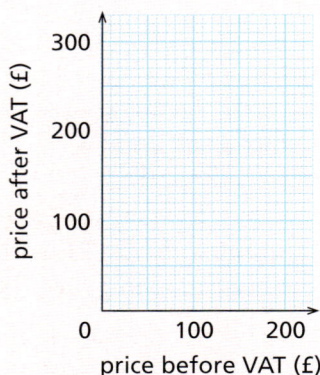

5 Given that 1 gallon is approximately 4.5 litres, prepare a conversion graph for gallons and litres.
6 Curtain material is sold 'by the metre'. One kind of material is advertised at £16 per metre. Construct a graph to convert between the length of the material (in metres) and the cost (in £).

Exercise 14.3 Ma1

Choose a country you would like to visit. Prepare graphs for the conversions you might need. These could include the following.

• A currency conversion
• A conversion between miles and kilometres (it might be helpful to mark the speed limits in the country you are visiting)
• A table converting between times in the 12 hour clock and times in the 24 hour clock (many countries make more use of the 24 hour clock than Britain)
• A table converting between British sizes and the local sizes (useful if you are thinking of buying clothes or shoes)

Lines not through the origin

So far, all the conversion graphs have gone through the origin. £0 is equal to 0 Swiss francs, or 0 $, or 0 ¥. If you have no money, you are broke in every currency! Some conversion graphs don't go through the origin. This happens if 0 in the first quantity doesn't correspond to 0 in the second quantity.

'genez Sauwe Hɛm'

'Hɛnan napu'

'pas d'argent'

'文 taし t̄='

'Kein Geld'

'non ho il denaro'

'geen geld'

'não tenho dinheiro'

Nowadays temperature is usually given in degrees Celsius, in which water freezes at 0 °C and boils at 100 °C. Many people prefer to use the Fahrenheit scale, in which water freezes at 32 °F and boils at 212 °F. Often recipe books and weather forecasts give temperatures in both scales.

A graph converting between Celsius and Fahrenheit does not go through the origin. as 0° Celsius corresponds to 32° Fahrenheit.

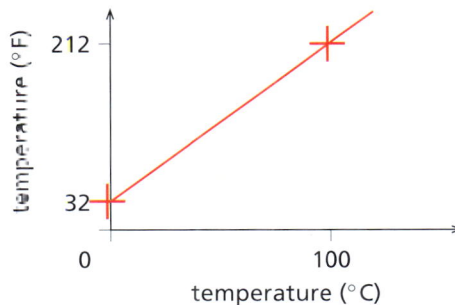

temperature (°F)

212

32

0   100

temperature (°C)

Gabriel Fahrenheit was a German physicist who made the first mercury thermometer, as well as inventing a temperature scale. In his scale 100 °F was about body temperature, and 0 °F was the lowest temperature he could get by freezing salty water.

Electricity bills usually include a *standing charge*. This means that you have to pay something, even if you used no electricity at all. Often repair people such as plumbers and electricians make a *call-out charge*, so that you have to pay something, even if they spend no time on the repair.

Example   The graph below shows what a plumber charges, in terms of the time he has to spend on the work.

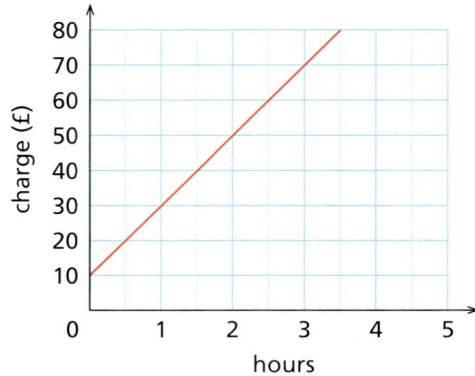

a What is his call-out charge?
b How much does he charge for a job taking 2 hours?
c A job costs £70. How long has it taken?

a Look at the graph. Even if the job takes 0 hours, the charge is still £10. The call-out charge is £10.
b From the graph, find that 2 hours on the *x*-axis corresponds to £50 on the *y*-axis.
   The charge for a 2 hours job is £50.
c From the graph, find that £70 on the *y*-axis corresponds to 3 hours on the *x*-axis.
   The job took 3 hours.

Exercise 14.4

1 The graph below shows telephone charges. The *x*-axis gives the number of units used, and the *y*-axis gives the charge.

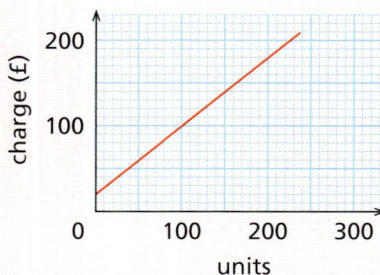

The standing charge is what you pay even if you don't use the telephone.

a How much is the standing charge?
b What is the charge if you use 150 units?
c A bill is £85. How many units were used?

2 The graph below shows the charges of a computer engineer, in terms of the amount of time she has to spend on a repair job.

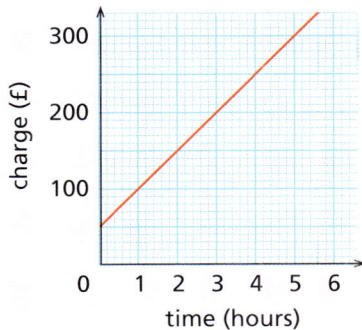

a What is her standing charge?
b What does she charge for a job taking 4 hours?
c She charges £190 for a job. How long did it take?

3 A bank will change pounds sterling into dollars ($), but it charges a commission. The graph below shows how many dollars you get for pounds.

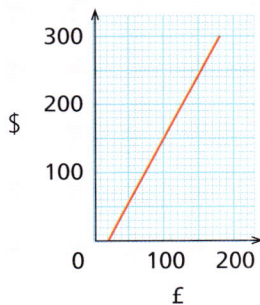

a How many dollars do you get for £50?
b How much do you have to pay to receive $200?
c How much is the commission?

You can construct a conversion graph between two quantities, provided that you know two pairs of values.

Example Construct a temperature graph converting between degrees Celsius and degrees Fahrenheit.

On the *x*-axis put the Celsius temperature, from 0 to 100. On the *y*-axis put the Fahrenheit temperature, from 0 to 250. The facts about these temperature scales are below.

Water freezes at $0\,°C$ and at $32\,°F$. So on the graph put a point at $(0, 32)$.
Water boils at $100\,°C$ and at $212\,°F$. So on the graph put a point at $(100, 212)$.

Join the points with a straight line. This graph can now be used to convert between the two scales.

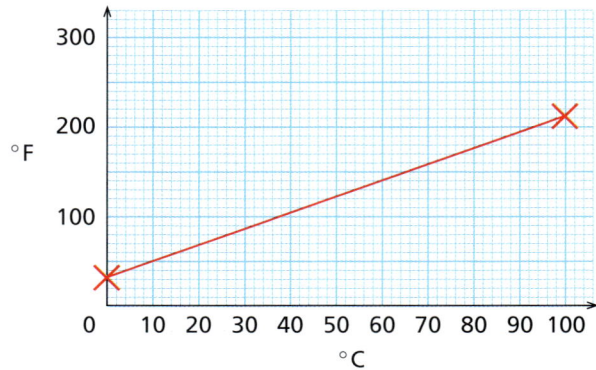

Exercise 14.5

1 A less common temperature scale is the Réaumur scale. On this scale, water freezes at 0 °R and boils at 80 °R. Copy the grid below onto graph paper, and make a conversion graph between Réaumur and Fahrenheit.

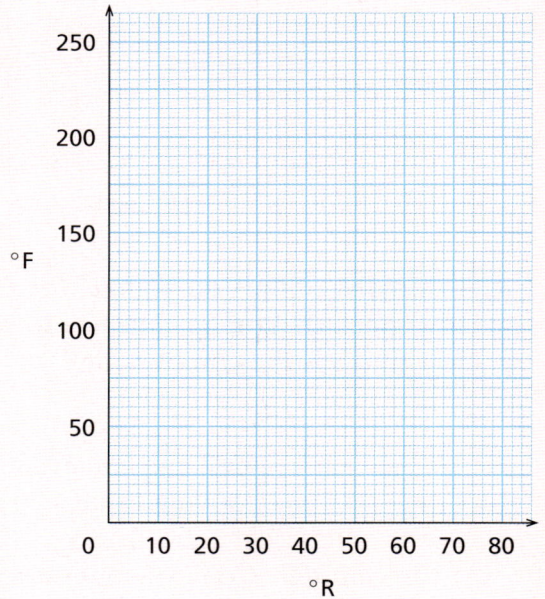

2 To hire a car from Ace Motorhire, you pay a fixed amount of £30 and a hire charge of £20 per day. Copy and complete this table.

number of days	1	2	3	4
charge (£)	50			110

Make a copy of the grid below, and use it to make a graph showing the cost of hiring for different numbers of days.

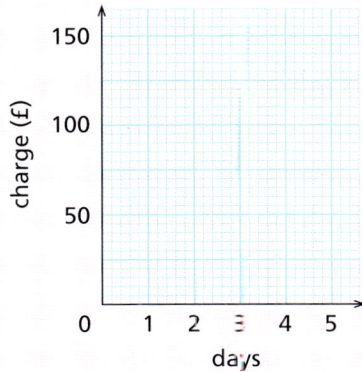

3 A bank will change £ to $, but charges £5 commission. After the commission has been paid, the bank will change at a rate of $1.60 per £. Copy and complete the table below.

£	5	15	25	35	45
$	0	16			

Make a copy of the grid below, and use it to make a graph showing the conversion between £ and $.

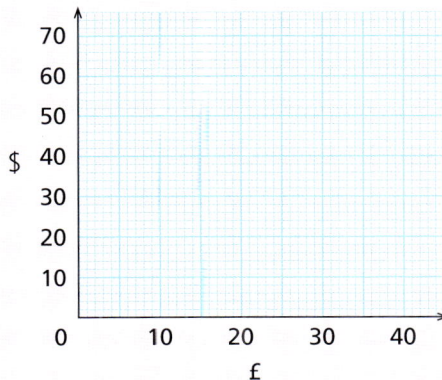

4 The graph of the earlier example converts between Celsius and Fahrenheit, but only up to 100 °C. For cooking, you need to have oven temperatures greater than that. For example, 250° Celsius is 482 °F. Copy the grid below and make a conversion graph that is useful for cooking temperatures.

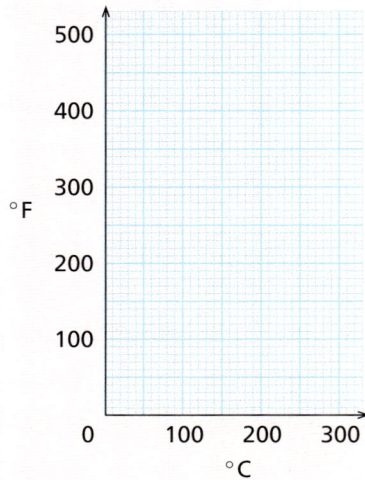

5 To roast a joint of pork until it is well done you set the oven for 180° and roast the meat for 35 minutes plus 35 minutes for every 500 grams of pork. Copy and complete the table below. Construct a graph to show how long you need to roast a joint.

weight (g)	500	1000	1500	2000
roasting time (minutes)				

Intersecting graphs

We can use conversion graphs to tell us useful facts. Suppose you want to hire a car. One company might have a large standing charge, but a low rate per day. Another might have a small standing charge, but a high rate per day. Which company do you use? If you have a graph showing the charges, you can use it to find out which company is better.

Example   To hire a van, I can use either Bonavans or Magimotors. The graph below shows how much each company charges for different numbers of days. If I want to hire the van for three days, which company should I use? When is it cheaper to hire from Bonavans?

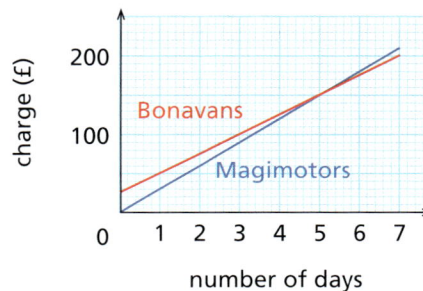

Look at the graph. For a three-day hire, Bonavans charges £100 and Magimotors charges £90. I should use Magimotors.
Try some other times.

For four days, Bonavans charges £125 and Magimotors charges £120.
For five days, Bonavans charges £150 and Magimotors charges £150.
For six days, Bonavans charges £175 and Magimotors charges £180.
For seven days, Bonavans charges £200 and Magimotors charges £210.

So, at five days, they charge equal amounts. After that, Bonavans is cheaper.
I should use Bonavans for more than five days' hire.

The 'break-even' point, where the companies charge the same amount, is at five days. We needn't do any calculation for this! Just look at the graph, and notice that the two lines cross at five days. Because the lines are crossing, the charges are the same at this point.

Example I can hire a car from two companies. Acehire charges £10 plus £30 per day. Zenith Cars charges £20 plus £25 per day. Copy and complete the table below for the charges of the two companies for different numbers of days. Plot graphs for the two companies. When is it cheaper to use Zenith Cars?

number of days	1	2	3	4
Acehire				
Zenith				

Do the Acehire figures first. For one day, I pay £10 plus £30, which comes to £40. For two days' hire I pay £10 plus 2 × £30, which comes to £70. The next figures are £100 and £130.
With Zenith Cars, for one day, I pay £20 plus £25, which comes to £45. For two days, I pay £20 plus 2 × £25, which comes to £70. The next figures are £95 and £120. The completed table is shown below.

number of days	1	2	3	4
Acehire	40	70	100	130
Zenith	45	70	95	120

Plot these points on a graph and join them up with straight lines.

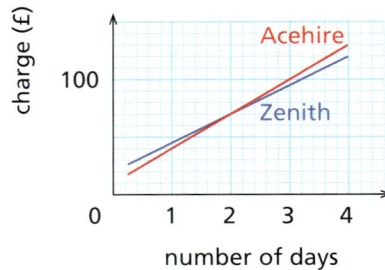

The graphs cross at (2, 70). After this point the Zenith graph is lower, so this company is cheaper. It is therefore cheaper to use Zenith Cars if you are hiring a car for more than two days.

Exercise 14.6

1 Alf and Wally are two plumbers. The graph below shows the amounts they charge, depending on the time taken. For a job lasting 2 hours, who should I call? When is it cheaper to use Alf rather than Wally?

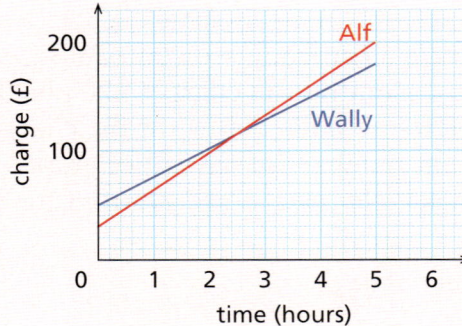

2 In my area, there is a choice between two electricity companies. The quarterly bills, for different amounts of electricity, are shown in the graph below. If I am going to use 200 units per quarter, which company should I sign up with? When is it better to use Wessex Electricity rather than Lyonesse Electricity?

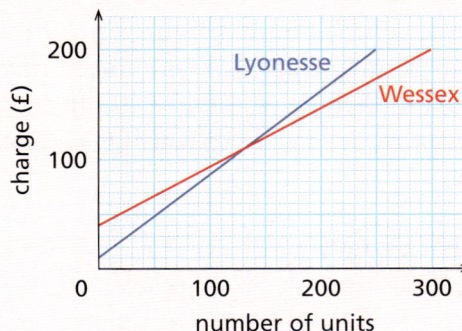

3 Two banks will change money from £ to $, but they both charge a commission. The graph below shows how many dollars you get for different amounts. To change £300, who should I go to? When is it better to use the Cardiff Bank rather than the Provincial?

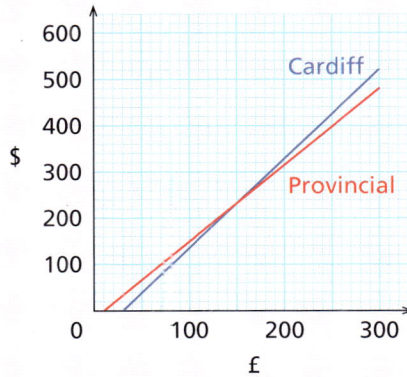

4 I can hire a wallpaper stripper from two do-it-yourself companies. Superbotch charges £10 plus £5 per day, and Papering Over the Cracks charges a fixed rate of £10 per day. Copy and complete the table below. Plot graphs on a copy of the grid below. When is it cheaper to use Superbotch?

number of days	1	2	3
Superbotch			
Papering Over the Cracks			

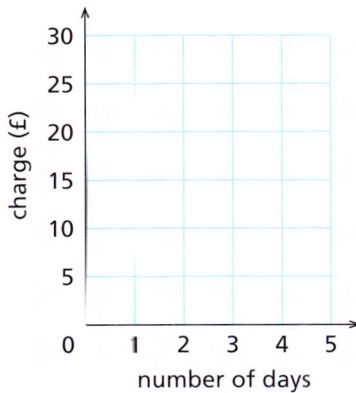

5 There is a choice of two removal companies for moving furniture. Pantechnicons R Us charges £200 plus £50 per 10 miles. Moveable Feast charges £220 plus £40 per 10 miles. Copy and complete the table below, and plot graphs on a copy of the grid. When is it cheaper to use Pantechnicons R Us?

number of miles	0	10	20	30
Pantechnicons R Us				
Moveable Feast				

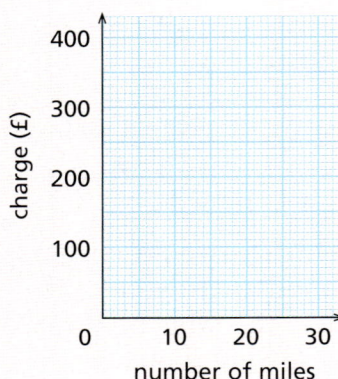

Travel graphs

The whole point of a graph is that it shows the connection between two quantities.

What happens when you go on a journey? At different times, your distance away from where you started is different. These two quantities, time and distance, can be shown on a graph.

A graph to show a journey, such as your daily journey to school, is a **travel graph**. Along the *x*-axis you put time, in minutes after leaving home. Up the *y*-axis you put distance, in kilometres or miles from your home. The graph then shows how far you have gone at different times.

If you travel to school at a steady rate, then your graph will be a straight line.

Example   Fiona leaves home and walks to school. The travel graph that follows shows her distance from home. How far away is the school? How long does she take to walk there?

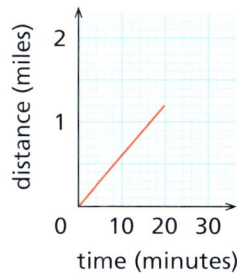

The end of the graph is at 20 minutes and 1.2 miles.
   The school is 1.2 miles away. She takes 20 minutes to walk there.

The graph in the example is a single straight line, so Fiona walks at a steady rate. If she stops at a shop to buy something, or if she has to go home to fetch something, then the graph will have more than one straight line.

Example   Ben walks to school, and stops at a shop on the way. The travel graph shows his distance from home. How long did he spend in the shop? How far is the shop from the school?

The horizontal part of the graph shows where his distance was not changing, that is, where he had stopped walking and was in the shop. The horizontal bit of the graph is between 15 minutes and 20 minutes.
   He spent 5 minutes in the shop.

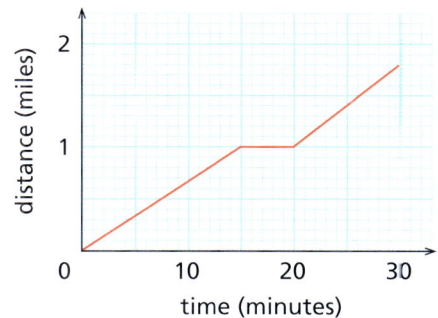

   The end of the graph (at school) is 1.8 miles from home. The horizontal bit of the graph (at the shop) is 1 mile from home. Subtract.
   The shop is 0.8 miles from the school.

Your journey might be more complicated, especially if it involves getting a bus or train as well as walking. The travel graph below is for someone who walks to the bus stop, waits for the bus, rides on the bus and then walks from the bus stop to the school.

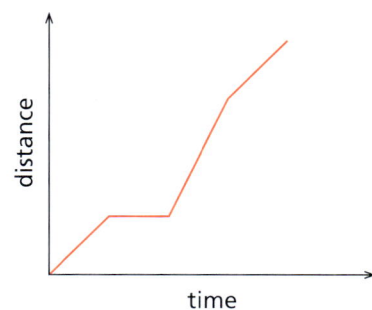

Exercise 14.7

1 The travel graph below shows Mr Joseph's journey to work.
   a How long does it take him?
   b How far is his work from his home?

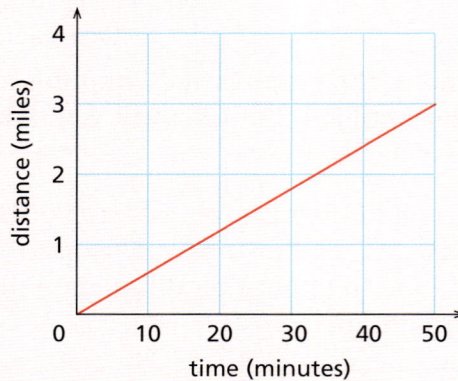

2 Norman walks to the library, returns a book and chooses another, then walks home. The diagram below is a travel graph for his journey.

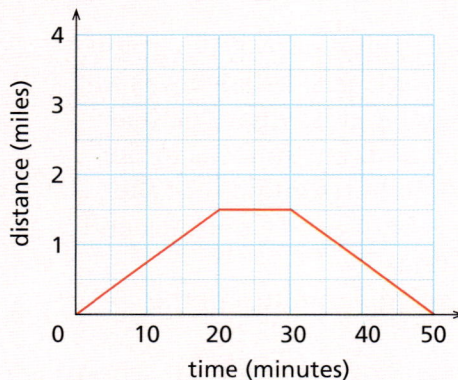

   a How long did he spend in the library?
   b How far is the library from his home?

3 Jessica walks home from school, stopping to talk to a friend. The travel graph below shows her distance from home.

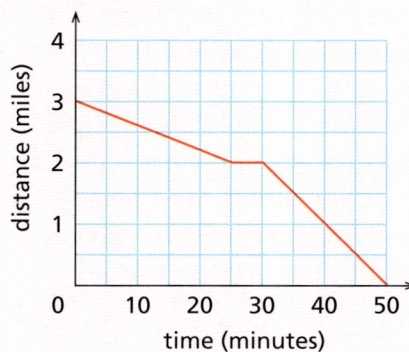

   a For how long was she talking to her friend?
   b When she was talking to her friend, how far from home was she?
4 Ms Knight travels from home to the office every day. She drives at a steady speed to the station car park, then she waits for a train. She takes the train to town and then walks to the office. Her journey is shown on the travel graph.

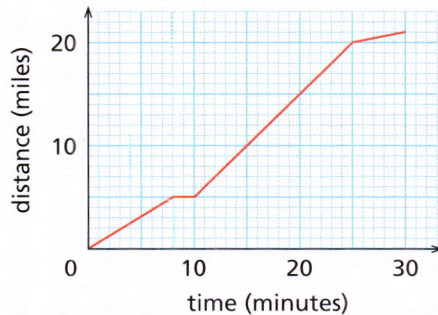

   a How far is it from home to the car park?
   b How long does she wait for a train?
   c If she left home at five minutes past eight, when does she arrive at the office?
5 Nicholas went on a cycle trip. His journey, consisting of three stages, is shown in the travel graph.

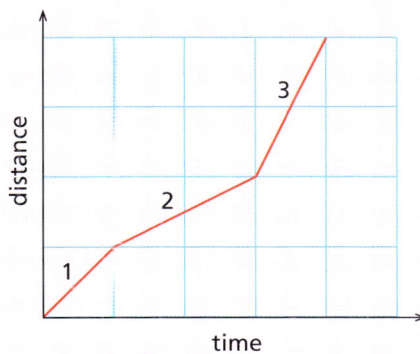

   a He was cycling uphill for one of the stages. Which stage was that?
   b One of the stages was downhill. Which stage was that?

Constructing travel graphs

You can construct a travel graph to show a particular journey. Put the time along the *x*-axis and the distance up the *y*-axis. Work out where the traveller is at various times, and draw straight lines on the graph corresponding to the different stages of the journey.

Examples   Nafisha walks to a shop. She covers 2 miles in 30 minutes, then spends 10 minutes shopping, then walks home in 40 minutes. Draw a travel graph to show Nafisha's journey.

She starts at home and after 30 minutes she is 2 miles away. Plot a point at (30, 2), then join up with a straight line.

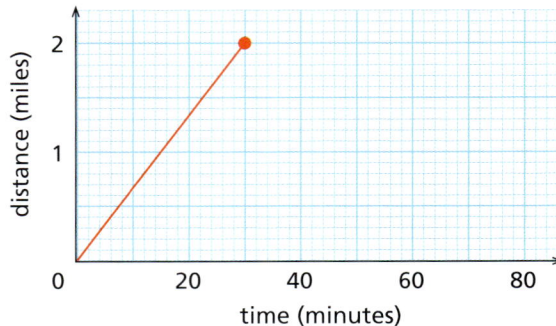

She stops for 10 minutes, so draw a horizontal line which represents 10 minutes.

She now goes home. She will reach home after a total of 80 minutes. Plot a point at (80, 0), then join up with a straight line. The completed travel graph is shown below.

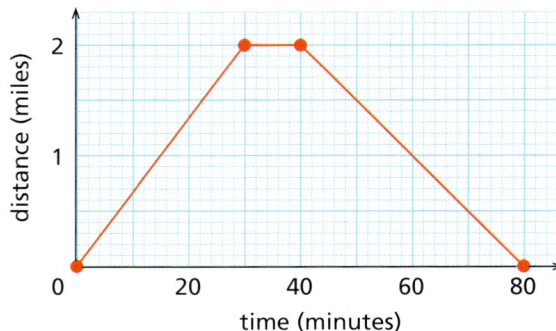

The Sutton family drive up the motorway, covering 50 miles in 1 hour at a steady speed. They stop at a service station for $\frac{1}{2}$ hour, then drive a further 120 miles in 2 hours, at a steady speed. Draw a travel graph to show their journey.

Show the first stage of the journey by drawing a straight line joining (0, 0) to (1, 50). Then add a horizontal line for the time at the service station, ending at $(1\frac{1}{2}, 50)$. Join the end of this line to the point $(3\frac{1}{2}, 170)$ for the final stage. The travel graph is shown below.

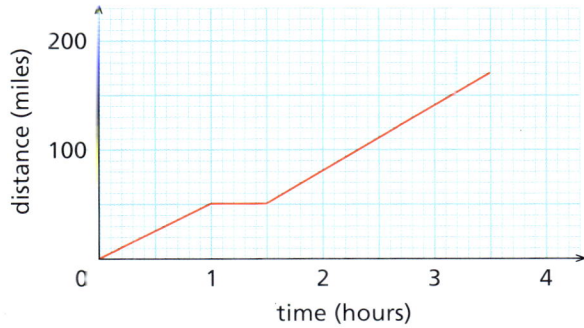

Exercise 14.8

1 Maria walks to visit her grandfather. He lives 3 miles away, and she gets there in 45 minutes. Copy the grid below on squared paper and draw a travel graph to show her journey.

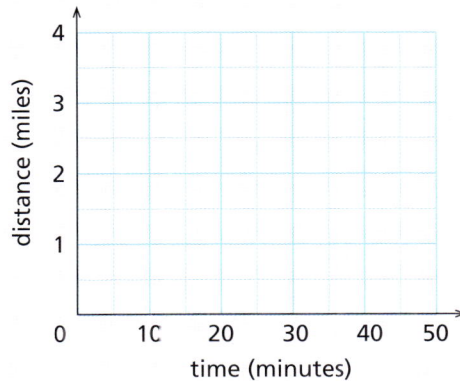

2 Mr Macpherson drives up a motorway. He covers 120 miles in 2 hours, at a steady speed. Draw a travel graph on a copy of the grid below to show his journey.

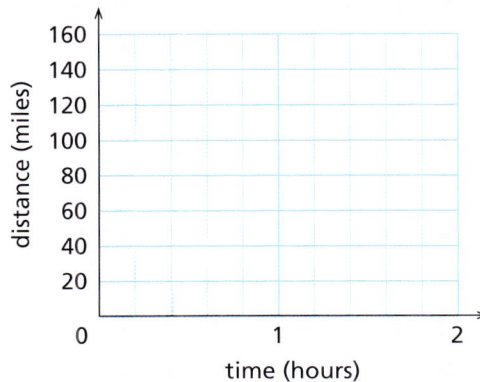

3 Jason rides a cycle to school. In 15 minutes he covers 3 miles. He then stops at a shop for five minutes, then he cycles the remaining 4 miles in 20 minutes. On a copy of the grid below, draw a travel graph to show his journey.

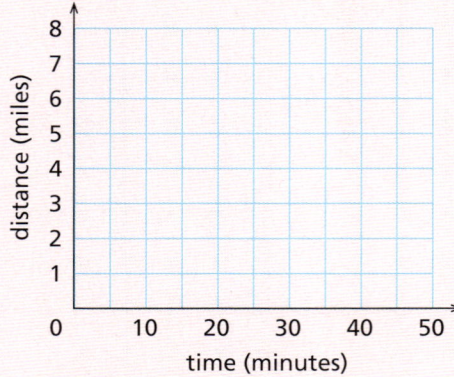

4 An aeroplane is flying from London to the Far East. It flies for 8 hours at a speed of 450 m.p.h. It stops for $1\frac{1}{2}$ hours to refuel, then covers the remaining 2000 miles at a steady speed in 5 hours. On a copy of the grid below, draw a travel graph to show the journey.

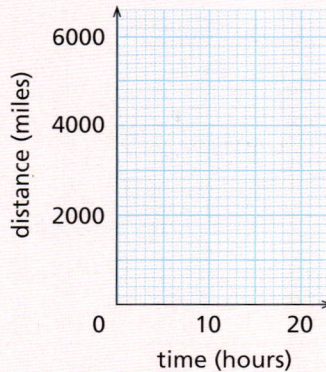

5 Jamie walks to school, covering 2 miles in 30 minutes. The moment he arrives he realises he has forgotten his homework. He rushes home, taking 20 minutes, then starts off for school again, also taking 20 minutes to get there. On a copy of the grid below, draw a travel graph to show his journey.

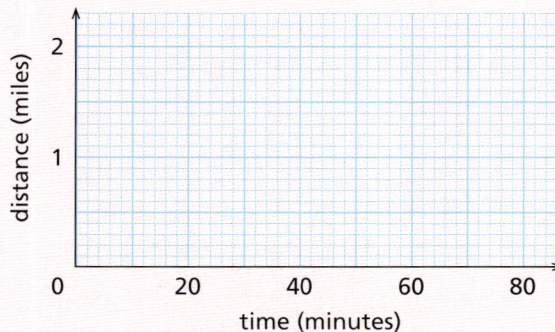

Exercise 14.9 (Ma1)

How do you get to school? Are you driven, or do you walk or cycle, or do you take a bus or train? Construct a travel graph to show your journey to school, or any journey that you do regularly.

SUMMARY

- You can make a **conversion graph** to convert between two quantities, such as two currencies.
- If there is a fixed rate of conversion, the graph is a straight line. You can use just two points to find the line, but a third point will provide a check that your line is correct.
- Sometimes the conversion graph will not go through the origin.
- Intersecting conversion graphs give us useful comparisons.
- A **travel graph** shows time along the x-axis and distance up the y-axis. You can use a travel graph to describe a journey.

Exercise 14A

1   The graph below shows the conversion between inches and centimetres. Use the graph to find
   a the length in centimetres of a 5 inch pencil
   b the length in inches of a 20 cm spoon.

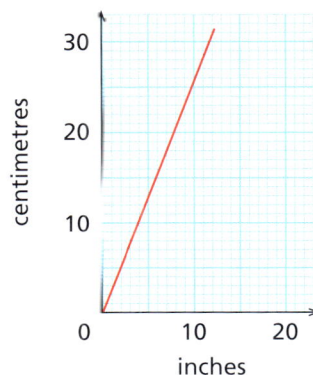

2 The Celsius and Réaumur temperature scales have the same freezing point of water, 0°, but the boiling point of water is 100° Celsius and 80° Réaumur. Make a copy of the grid below, and on it construct a conversion graph for the two scales.

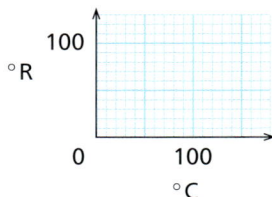

°R
100

0   100

°C

3 The graph below shows the charges of a roofer, which depend on how long the job takes.

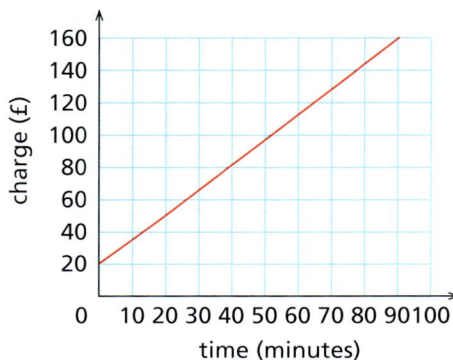

charge (£)

160
140
120
100
80
60
40
20

0   10 20 30 40 50 60 70 80 90 100

time (minutes)

   a What is his call-out charge?
   b What will he charge for work lasting 20 minutes?
   c If he charges £120, how long did the work last?

4 Andrew is going on a sponsored run. His mother promises him £4 just for starting, and £1 for each mile he runs. On a copy of the grid below draw a graph of the amount he will receive from her.

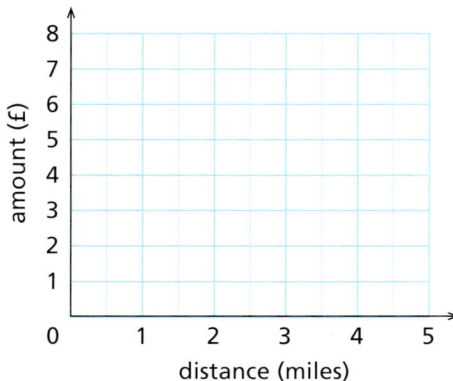

amount (£)

8
7
6
5
4
3
2
1

0   1   2   3   4   5

distance (miles)

5 I need to get my front door locks repaired. Lock 'n' Key charge a fixed rate of £30 per hour. Safe as Houses charge a £10 call-out fee, then £25 per hour. On a copy of the grid below, draw graphs showing their charges. When is it better for me to use Lock 'n' Key?

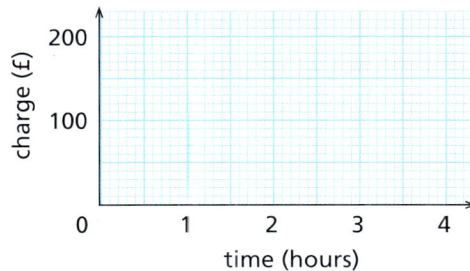

6 The travel graph below shows the Johnson family's journey along the motorway, during which they stopped at the service station.

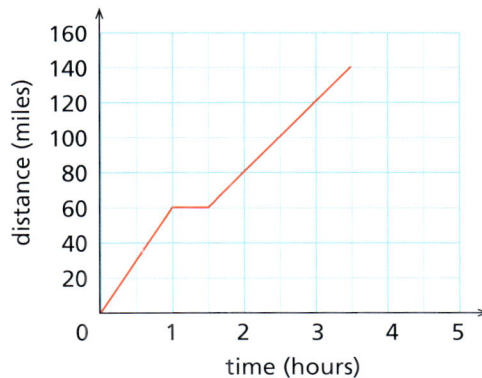

  a What distance did they travel?
  b How long did they stop at the service station?
  c How long did the journey take after leaving the service station?
7 Ms Langton drives to work at a steady speed, taking 40 minutes for the 30 mile journey. On a copy of the grid below draw a travel graph to show her journey.

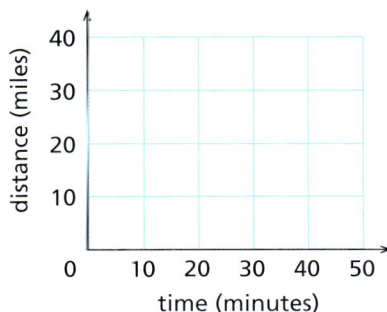

8 A balloon is flying 1000 m above the ground. At 12 o'clock the balloon starts to descend at 10 m per second. At the same time another balloon starts to rise from the ground at 5 m per second. On a copy of the grid on the right draw graphs showing the height of each balloon above the ground. When do they pass each other? How high are they when this happens?

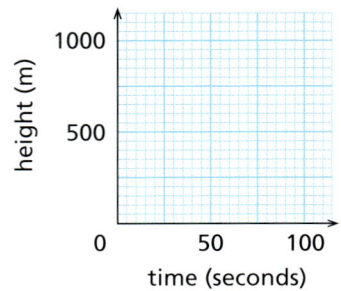

Exercise 14B

1 The graph on the right converts between US dollars and Canadian dollars.

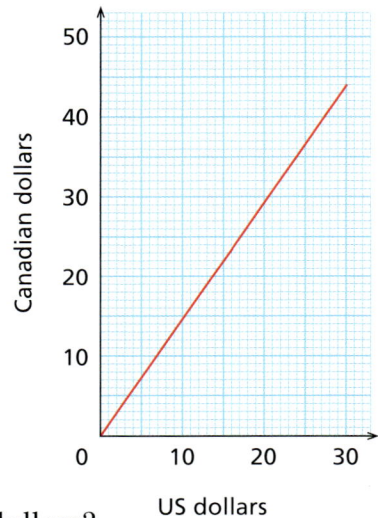

a How much are 40 Canadian dollars worth in US dollars?
b How much are 30 US dollars worth in Canadian dollars?

2 A bank will change money between £ and $, but it has different rates for buying and selling. The red line of the graph on the right is for changing £ into $, and the blue line is for changing $ into pounds.

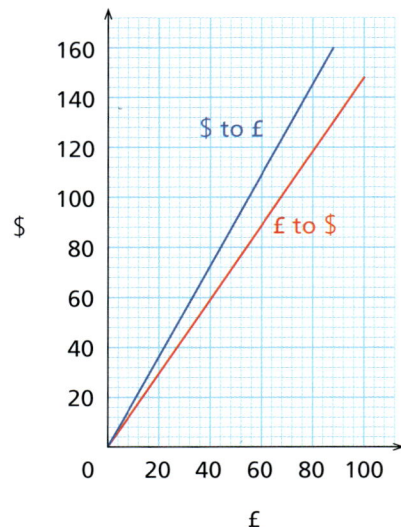

a How much in $ do I get for £60?
b How much in £ do I get for $50?
c I change £80 into $, then I change my mind and change the dollars back into pounds. How much have I lost?

3 A petrol station sells unleaded petrol at 70 pence per litre. Make a copy of the grid below, and use it to construct a conversion graph between the number of litres bought and the cost in £.

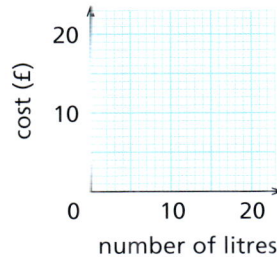

4 In the Bible, distances are often given in *cubits*. The cubit is the length between the elbow and the tip of the middle finger for an adult man. One cubit is about 45 cm or $1\frac{1}{2}$ feet. Make a conversion graph between feet and cubits.

5 A solicitor offers her services for a fixed fee plus a charge for each $\frac{1}{4}$ hour. The graph below shows her scale of charges.

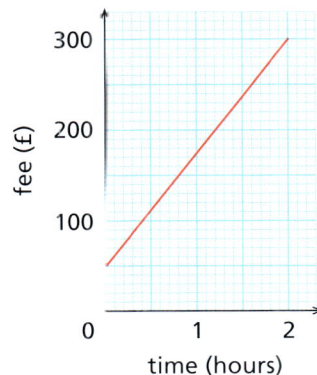

a What is the fixed fee?
b What is the total cost for a client after 1 hour?
c What is the rate per $\frac{1}{4}$ hour?

6 A bank takes £2 commission and offers Thai currency at a rate of 75 baht to the £. Copy and complete the table below. On a copy of the grid below construct a conversion graph between £ and baht.

amount (£)	2	10	20	30
amount (baht)	0	600		

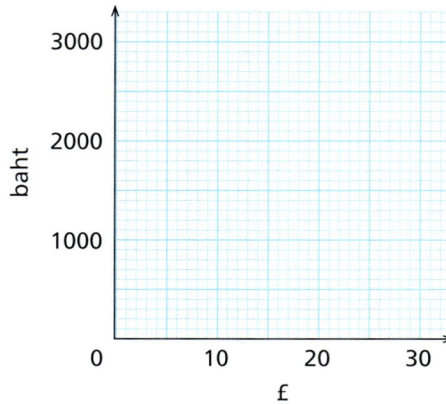

7 Tamsin sets out for her favourite ice cream shop. She walks briskly to the sea front, walks slowly along the sea front, then walks quickly for the final stage of the journey. The graph on the right shows her journey.

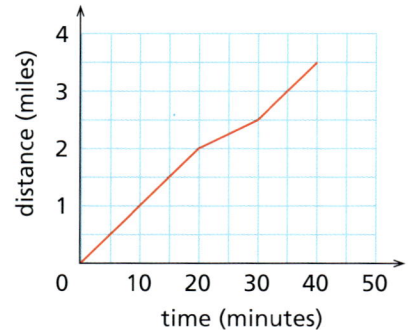

a How long did the first stage take her?
b How far did she walk along the sea front?
c How far away is the ice cream shop?

8 Look at the graph below. Explain why it cannot be a travel graph (except in science fiction!).

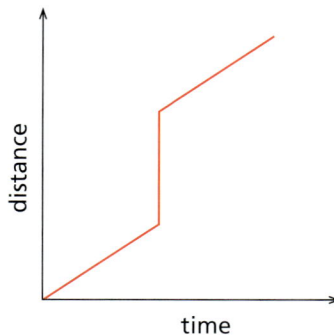

Exercise 14C Ma1

More and more people are getting connected to the Internet! You need a computer for this, of course, and an extra piece of hardware called a modem. You also need to sign up with an organisation called an *Internet service provider* (ISP). This provider will enable you to send and receive e-mail, to browse the net, and so on.

Most ISPs make a monthly charge. Some have a fixed rate per month, regardless of how often you use them. Some have a lower fixed rate, and charge you for each hour you spend using their services.

Find out the rates of various ISPs. (Some examples of ISPs are Compuserve, AOL, Demon, LineOne.) Make a large graph showing how much they charge for different levels of usage. Which would be cheapest for you?

Exercise 14D Ma1

In this chapter the travel stories have all been given to you! Now is the time to use your imagination! Below are some travel graphs – try to think up a story involving travel which they fit.

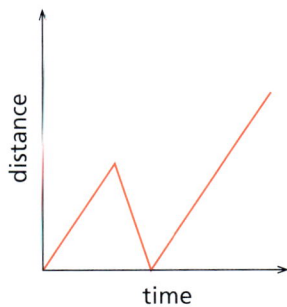

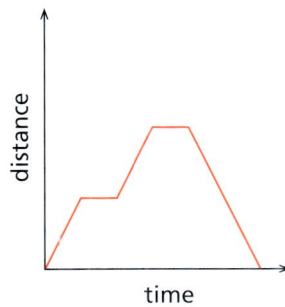

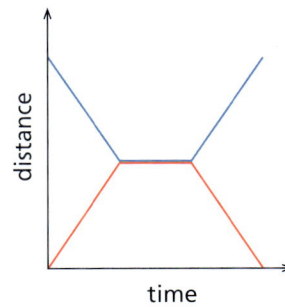

15 Averages

Isabel and Kathleen are both keen cross-country runners. They have each run over the same course eight times. Isabel's times were (in minutes)

45 48 44 46 48 50 47 48

Kathleen's times, in minutes, were

43 42 41 40 42 43 88 39

They argue about who is better.

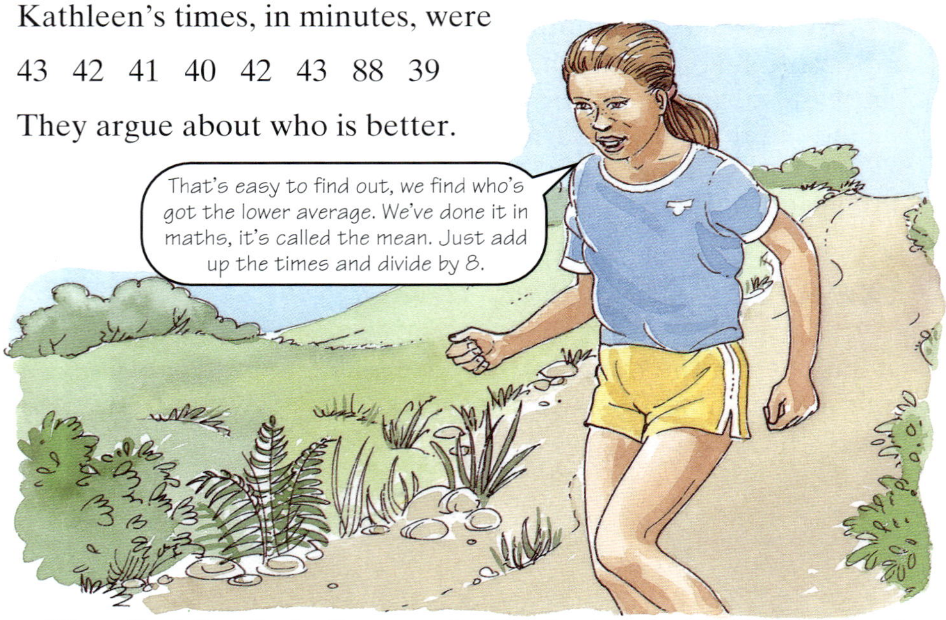

> That's easy to find out, we find who's got the lower average. We've done it in maths, it's called the mean. Just add up the times and divide by 8.

They find that Isabel's mean is 47 minutes, and that Kathleen's mean is 47.25 minutes.

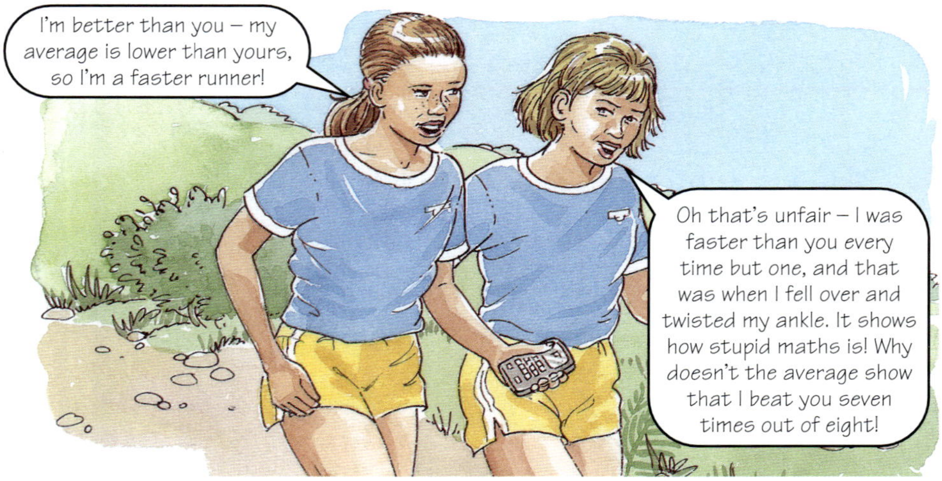

> I'm better than you – my average is lower than yours, so I'm a faster runner!

> Oh that's unfair – I was faster than you every time but one, and that was when I fell over and twisted my ankle. It shows how stupid maths is! Why doesn't the average show that I beat you seven times out of eight!

The mean may be one of the original numbers, or it may not. Isabel's mean time was 47 minutes, which was one of her times. Kathleen's mean was 47.25, which wasn't one of her times.

Mean

You have already found the **mean** of sets of numbers. The mean is what we normally think of when we say 'average'. To find the mean of 10 numbers, add them up and divide by 10.

Exercise 15.1

(Mainly revision)
In each of questions 1 to 3 find the mean of the set of numbers.

1 3, 6, 3, 4, 9, 4, 3, 1, 6, 7

2 −2, −2, −1, −1, 0, 0, 1, 1, 1, 1, 2, 2, 2, 3, 5

3 1.43, 1.32, 1.74, 1.57, 1.33, 1.50, 1.42, 1.49, 1.35

4 Ten pupils took a test and gained the marks below. What was the mean mark?

    14 18 20 8 13 10 18 19 9 14

5 The midday temperatures, in °C, of twelve summer days are given below. Find the mean temperature.

    26 19 27 24 26 20 22 24 28 20 21 23

6 The level of a reservoir above or below its original height is measured over ten days. The results, in inches, are given below. Find the mean level.

    3 2 0 −1 −3 −5 −6 −4 −1 3

7 Janine and Carrie both keep records of how much they spend each week. The results, in £, for nine weeks are given below. Find the mean amount for each girl. Who spends more?

Janine	15.70	16.35	18.24	18.20	22.55	20.12	14.66	15.20	14.33
Carrie	10.35	14.26	9.05	12.40	16.32	18.30	16.80	12.47	8.97

8 Every morning Mr and Mrs Robinson find how long it takes them to drive to work. The results (in minutes) for ten mornings are given below. Find the mean time for each person. Who spends longer driving to work?

Mr Robinson	28	21	23	24	20	25	32	26	20	22
Mrs Robinson	24	31	27	26	25	26	28	26	27	24

9 There will be ten maths tests this term. Jason's marks in the first nine tests are below. What must he get in the tenth test to ensure that his mean mark is at least 60?

    58 63 55 68 55 52 62 66 58

10 Jean is training for the 100 m sprint. The times, in seconds, of her first five attempts are below. What must she achieve in the sixth attempt if her mean time is to be less than 11 seconds?

    10.9 11.3 11.0 11.1 10.8

Median

Suppose you have a group of ten people, and you want to know their average income. Perhaps you find that their mean income is £50 000 per year. So these people are fairly well off, aren't they?

They might be. But it isn't obvious. Perhaps the incomes are

£8800 £7600 £6900 £9300 £8500 £6800 £6700 £8100 £9800
£427 500

Of these people, all but one earns less than £10 000! It's very misleading to say that the average income is £50 000, as this is far more than the income of all but one of the people. The reason that the mean is so high is that it is distorted by the one person with a very large income, perhaps a football star or the director of a big company. We need another way to show the average income of this group.

Exercise 15.2

1 Ten people run an 800 m race. Their times, in seconds, are given below. What is the mean time? Explain why the mean does not give a good idea of the average time.

210 205 199 213 968 198 208 210 208 211

2 A college calls eight students for interview. The times they spend in the interview room, in minutes, are given below. What is the mean time? Explain why the mean does not give a good idea of the average time.

15 14 13 15 15 16 1 15

In the example about incomes, and in the questions of exercise 15.2, the mean is distorted by a single very large value or very small value. For data sets like these, with a single extreme value, we need an average which tells us what the 'middle' value is. We need an average for which half the data are less than it and half are more than it. This average is the **median**.

● *To find the median of a set of data, arrange the data in increasing order. If there is an even number of data, the median is half-way between the two middle values. If there is an odd number of data, the median is the middle value.*

Examples Find the median income of the ten people mentioned on page 274. Their incomes were:

£8800   £7600   £6900   £9300   £8500   £6800   £6700   £8100
£9800   £427 500

Arrange the incomes in increasing order, as shown below.

£6700   £6800   £6900   £7600   £8100   £8500   £8800   £9300
£9800   £427 500

Now we can split the group into two halves: the richer half and the poorer half. The poorer half earn between £6700 and £8100, and the richer half earn between £8500 and £427 500. Take half-way between £8100 and £8500, at £8300.
   Half the group earn less than £8300 and half earn more.
   The median income is £8300.

A group of eleven pupils took a test, and their marks are given below. Find the median mark.

38   35   39   40   37   34   31   32   0   33   37

Arrange these marks in increasing order.

0   31   32   33   34   35   37   37   38   39   40
                          ↑

The middle number is the 6th number, which is 35. This is the median.
   The median mark is 35.

In the first example, the median gives a much better idea of the average income than the mean because the median is closer to what most people in the group earn. In the second example the median gives a much better picture of the class as a whole than the mean would. The mean, 29.3, is dragged down by a single low score. (Perhaps that person was ill, or fell asleep during the test!)

Exercise 15.3

In each of questions 1 to 5 find the median of the data set.

1 13, 12, 19, 14, 15, 12, 18, 16, 12, 11
2 34, 35, 38, 33, 41, 30, 32, 31, 29
3 193, 199, 204, 188, 194, 196, 194, 206, 200, 193, 198, 202
4 2.45, 2.94, 2.33, 2.42, 2.19, 2.50, 2.44, 2.48, 2.43, 2.35
5 −6, −3, 4, 0, 4, 1, 1, −4, −2, −2, −1
6 Find the median of the race times given in question 1 of exercise 15.2. They were:

210 205 199 213 968 198 208 210 208 211

What might be a reason for the exceptional value?

7 Find the median of the interview times given in question 2 of exercise 15.2. They were:

15 14 13 15 15 16 1 15

What might be a reason for the exceptional value?

8 The temperatures in an office were found on nine occasions. The results, in °C, are given below. Find the median temperature. What is a possible reason for the exceptional value?

23 24 25 22 23 24 10 24 23

9 Look again at the conversation at the beginning of this chapter. Find the median time for both Isabel and Kathleen. Who do you think is the better runner?

10 Asmat and Nick are the top two in their maths set. In seven tests their scores were:

Asmat   43   46   50   48   3   44   46
Nick   42   44   47   48   46   43   44

For each boy, find the mean score and the median score. Who do you think is better at maths?

Mode

You follow the fortunes of a football team. In 12 matches they scored the following numbers of goals.

0 2 4 1 0 0 1 0 0 3 4 0

What is the average number of goals? How many goals are you most likely to see when you go to watch them play?

The mean number of goals is $1\frac{1}{4}$. You won't be able to see the mean number of goals. You can't have $\frac{1}{4}$ of a goal!

Put the scores in order, as

0 0 0 0 0 0 1 1 2 3 4 4

The median number of goals is half-way between 0 and 1, so the median is $\frac{1}{2}$. You won't be able to see the median number of goals, either, as you can't score half a goal!

The most likely number of goals is 0. This is the **mode**, or the **modal** score.

The mode of a set of data is the most frequently occurring number. You don't have to do any arithmetic to find the mode, just spot which number occurs most often.

Example Twelve children were asked how many brothers or sisters they had. The result is below. What is the mode?

0 1 1 2 1 3 1 4 0 0 5 2

There are three 0s, four 1s, two 2s, a 3, a 4 and a 5. The most common is 1.

The mode is 1.

The mode is the number which occurs most often. There may be two or more numbers which occur with the same frequency. Look at this data.

0 0 0 1 1 2 2 2 3 4

The most frequent numbers are 0 and 2, both of which occur three times. Here there are two modes.

If all the numbers are different, the frequency of each number is 1. Then it isn't sensible to use the mode at all!

Exercise 15.4

In each of questions 1 to 5 find the mode of the set of data. If there is more than one mode give them all.

1 1, 2, 3, 1, 2, 4, 8, 1, 2, 3, 1
2 10, 12, 13, 11, 14, 11, 12, 13, 14, 11, 10, 11
3 −2, −4, 0, 1, 1, −1, −1, 1, 0, 0, −3, 0, 2
4 5, 6, 2, 3, 9, 10, 4, 6, 11
5 27, 33, 25, 38, 29, 21, 25, 35, 38
6 The numbers of goals scored by a hockey team in nine matches are given below. Find the modal score.

0 2 1 3 1 4 1 0 1

7 A group of 10 pupils took a test, with a mark out of 10. The results are given below. What is the modal mark?

9 10 8 1 10 5 10 9 7 8

8 A group of seven people were asked for their shoe size. The results are given below. Find the modal size.

8 8½ 7½ 7 7 8 8

9 The scores of a basketball team over ten matches are given below. Is it sensible to use a modal score?

33 65 20 41 55 57 51 39 44 49

10 The salaries of six employees are given below. Is it sensible to use a modal salary?

£23 410 £19 340 £14 290 £16 500 £20 350 £19 660

Mean from a frequency table

Suppose the data you get is already sorted into a frequency table. You can still find the mean.

Example   A firm received 30 letters with first class stamps from different places. The table below gives the number of days that the letters had taken to arrive. Find the mean number of days.

number of days	1	2	3	4	5
frequency	12	8	6	3	1

We can write out all the data in full. The letter took 1 day on twelve occasions, so there are twelve 1s. Then there are eight 2s, six 3s, three 4s and one 5. The data are

1, 1, 1, 1, 1, 1, 1, 1, 1, 1, 1, 1, 2, 2, 2, 2, 2, 2, 2, 2, 3, 3, 3, 3, 3, 3, 4, 4, 4, 5

There are 30 numbers here, and their total is 63. Divide 63 by 30, obtaining 2.1.
   The mean number of days is 2.1.

Exercise 15.5

1 The frequency table below gives the numbers of brothers or sisters of 25 children. Write out the numbers in full, and find the mean number.

number of brothers/sisters	0	1	2	3	4
frequency	8	9	4	3	1

2 The frequency table below gives the numbers of times 30 motorists took the driving test before they passed. Write out the numbers in full, and find the mean number.

number of tests	1	2	3	4	5
frequency	14	8	5	2	1

When you looked at the example, and when you did the last exercise, did you notice a quicker way? You don't have to write out all those numbers in full! Look again at the example.

12 letters took 1 day each.
This gave a total of 1 × 12 days.   12 days
8 letters took 2 days each.
This gave a total of 2 × 8 days.   16 days
6 letters took 3 days each.
This gave a total of 3 × 6 days.   18 days
3 letters took 4 days each.
This gave a total of 4 × 3 days.   12 days
1 letter took 5 days.
This gave a total of 5 × 1 days.   5 days

Add all these totals, obtaining, as before, 63 days. There were 30 letters, so divide 63 by 30.

● *To find the mean from a frequency table, multiply each value by its frequency. Add these, then divide by the sum of the frequencies.*

$$\text{mean} = \frac{\text{sum of (values} \times \text{frequencies)}}{\text{sum of frequencies}}$$

Example   A group of 20 men were asked how many children they had. The results are in the table below. Find the mean number of children.

number of children	0	1	2	3	4
frequency	3	8	4	3	2

Follow the method above. Multiply each number by its frequency. Add these, then divide by the sum of the frequencies.

$$\frac{0 \times 3 + 1 \times 8 + 2 \times 4 + 3 \times 3 + 4 \times 2}{3 + 8 + 4 + 3 + 2} =$$

$$\frac{0 + 8 + 8 + 9 + 8}{20} = \frac{33}{20} = 1.65$$

The sum of frequencies is just the total number of people or items in the sample.

So the mean number of children is 1.65.

Exercise 15.6

1 This table comes from question 1 of exercise 15.5. Find the mean number of brothers or sisters using the method above.

number of brothers/sisters	0	1	2	3	4
frequency	8	9	4	3	1

2 This table comes from question 2 of exercise 15.5. Find the mean number of driving tests using the method above.

number of tests	1	2	3	4	5
frequency	14	8	5	2	1

3 A test is taken by 40 pupils. The results are in the frequency table below. Find the mean mark.

mark	6	7	8	9	10
frequency	3	8	19	8	2

4 A survey is made into the number of times per day people eat meat. The results are given below. Find the mean number of times.

number of times	0	1	2	3
frequency	23	38	26	13

5 People are asked how often they have been to the cinema in the last month. The frequency table below gives the results. Find the mean number of visits.

number of visits	0	1	2	3	4
frequency	14	36	19	14	10

Median from a frequency table

Suppose data is given in a frequency table of values. We can find the median from it.

Look again at the example on page 278, about the time 30 first class letters took to reach their destination. When the times were arranged in order, they were

1, 1, 1, 1, 1, 1, 1, 1, 1, 1, 1, 1, 2, 2, 2, 2, 2, 2, 2, 2, 3, 3, 3, 3, 3, 3, 4, 4, 4, 5

As there are 30 letters, the median time is half-way between the times of the 15th and the 16th letters. Both of these numbers are 2, hence the median time is 2 days.

We can find the median from a frequency table without having to write out all the data in order. For this example, add up the frequencies until we have included the 15th and 16th values. There are twelve 1s, eight 2s, six 3s, three 4s and a 5. Adding the 1s and 2s together, we have 20 in all, so the 15th and 16th values must lie in the 2s. Both values must be 2.

Here's how to find which value to take.

> If you have 23 values, halve 23 to get 11.5. Add 0.5, getting 12. Find which frequency group the 12th value lies in. This gives you the median.

To find the median from a frequency table,

- *if you have an odd number of values, halve the number of values and add 0.5. This numbered value is the median*
- *if you have an even number of values, halve the number of values. The median is midway between this numbered value and the next.*

Example   The midday temperature was measured for the 31 days of August. The results are given below. What was the median temperature?

temperature (°C)	20	21	22	23	24	25	26	27
frequency	1	2	5	10	9	3	0	1

There are 31 days, which is an odd number. Half of 31 is 15.5. Add 0.5, obtaining 16. So the median temperature is the 16th. To find the 16th value, we don't have to write out all the numbers in order. Just count up the frequencies until you include 16.

$$1 + 2 = 3 \qquad 1 + 2 + 5 = 8 \qquad 1 + 2 + 5 + 10 = 18$$

We have passed 16. The 16th temperature must be in the 23° group. The median temperature is 23 °C.

Exercise 15.7

1 A group of 40 pupils took a test, with the following results. Find the median mark.

mark	0	1	2	3	4	5	6	7	8	9	10
frequency	0	0	2	5	6	10	6	5	2	0	4

2 A group of 53 children were asked how many brothers or sisters they had. Find the median number of brothers or sisters.

number	0	1	2	3	4	5	6
frequency	13	17	10	8	3	1	1

3 This frequency table is from question 3 of exercise 15.6. Find the median mark.

mark	6	7	8	9	10
frequency	3	8	19	8	2

4 This frequency table is from question 4 of exercise 15.6. Find the median number of times.

number	0	1	2	3
frequency	23	38	26	13

5 This frequency table is from question 5 of exercise 15.6. Find the median number of visits.

number	0	1	2	3	4
frequency	41	36	19	14	10

Mode from a frequency table

All the work has been done for you!

It is easy to find the mode from data in a frequency table. Just look to see which is the largest frequency – that gives you the mode!

Example   What is the modal temperature for the example on page 281?

temperature (°C)	20	21	22	23	24	25	26	27
frequency	1	2	5	10	9	3	0	1

Look along the frequency row. The largest frequency is 10.
   So the modal temperature is 23°.

Exercise 15.8

For questions 1 to 5 in exercise 15.7, write down the mode.

SUMMARY

- To find the **mean** of *n* numbers, add them up and divide by *n*. The mean can be distorted by a single extreme value.
- To find the **median** of a set of values, arrange them in increasing order. If there is an odd number of values, take the middle value. If there is an even number of values, take half-way between the middle two values.
- The **mode** of a set of numbers is the most common number. There may be more than one mode.
- To find the mean from a frequency table, multiply each value by its frequency. Add, then divide by the sum of the frequencies.
- To find the median from a frequency table, add the frequencies until you have reached the half-way mark.
- To find the mode from a frequency table, pick the value with the largest frequency.

Exercise 15A

1 Find the mean of these numbers.

34 91 54 58 22 41 54 88 74

2 On ten days the temperature of an air-conditioned room was found. The results, in °F, are given below. Find the mean temperature. Why is the mean temperature not a good guide for the average temperature?

75 73 74 72 93 74 71 77 74 75

3 Find the median of the temperatures given in question 2. What is a possible reason for the exceptional value?

4 Bricket and Botchit are two small construction companies, each with eight employees. Their absentee records show how many days in a year each employee was absent. Find the mean and median for both companies. Which company do you think has the worse absentee record?

Bricket	2	3	5	2	6	5	4	0
Botchit	0	1	0	1	1	17	2	1

5 Find the mode of the following numbers.

4 2 6 2 3 2 1 1 3

6 Find the mean of the frequency table below.

number	1	2	3	4	5
frequency	8	12	6	3	1

7 A music store keeps a record of the daily sales of a certain CD. The results are below. Find the mean number of copies.

number of copies	0	1	2	3	4	5	6
frequency	1	4	8	5	3	0	1

8 For the data in question 7, find the median number of copies sold.
9 For the data in question 7, write down the mode of the number of copies sold.
10 A politician says that more than half the workers in this country earn less than the average income. What average is he talking about?

Exercise 15B

1 The numbers below are the marks of 10 students in an exam. Find the mean mark.

66 71 59 60 88 41 49 63 55 50

2 Twelve people took an exam. The times, in minutes, they spent in the exam room are given below. Find the mean time. Why is the mean time not a good guide for the average time?

91 89 90 90 88 2 87 85 92 88 84 86

3 Find the median of the times in question 2. What is a possible reason for the exceptional value?

4 Over nine one-day cricket matches, the scores of two batsmen were as follows. For each batsman, find the mean and the median. Who do you think is the better batsman?

Jeeves	10	0	79	7	107	8	13	19	0
Sutcliffe	31	48	0	6	35	38	43	17	24

5 Find the mode of the times in question 2.
6 The frequency table below gives the number of foreign countries visited by 40 people. Find the mean number of countries.

number	0	1	2	3	4	5
frequency	6	13	10	6	4	1

7 In question 6, find the median number of countries.
8 In question 6, write down the mode of the number of countries.

9 Which is larger, the mean income of people in the UK or the median income?

10 It is said that in 1947 the average number of pairs of trousers per man was 1. Which average is meant?

Exercise 15C   Ma1

To measure how self-centred a person is, you can count how often he or she uses the words 'I', 'me' and 'mine'. Find newspaper or magazine interviews with two famous people that you are interested in, and count the number of uses of these words. Who is more self-centred?

Exercise 15D   Ma1

You can use averages to follow the progress of a sports team. For several football teams, keep records of

- the number of goals the team scores in each match
- the number of goals scored *against* the team in each match

For each team, find the mean, median and mode of these numbers. Do these averages tell you where the team is in the league? In other words, is the team with the highest average also the highest in the league?

16 Area

The area of a surface tells us how large it is. If we are painting a room, we need to know how much paint will cover the walls and ceiling. If we are tiling a bathroom wall, we need to know how many tiles to use.

One standard size of tile is 6" by 6".

6" means 6 inches.

If we are lucky, we may be able to cover a wall with whole tiles. This will happen if the height and width, in inches, of the wall we want to cover are both divisible by 6.

This is unlikely to happen. Almost certainly, we will have to cut tiles in order to cover the surface. We will have to use fractions of a tile, and the area will also be fractional.

What if the wall isn't a rectangle? We will have to make a slanting cut along the tile, and it will be harder to find the area. We will need to know the area of a triangle, as well as that of a rectangle.

Area of a rectangle

In Book 1 you worked out the area of rectangular shapes. You found how many 'standard tiles' would fit into a rectangle. For small areas, a standard tile could be a square of side 1 cm. This has area 1 cm^2.

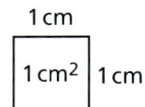

1 cm

| 1 cm^2 | 1 cm |

This diagram shows a rectangle which is 4 cm wide and 3 cm high. There are 12 standard 1 cm^2 tiles which fit into the rectangle. Hence the area of the rectangle is 12 cm^2.

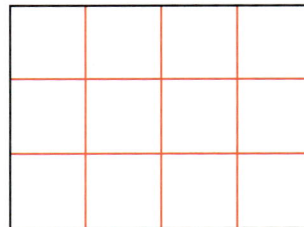

Of course, you do not need to count the tiles one by one. Once you have found a formula, use it! You can just multiply together the width and height of the rectangle.

$$\text{area} = 4 \text{ cm} \times 3 \text{ cm} = 12 \text{ cm}^2$$

Generally,

Width and height are often called length and breadth.

- *area of a rectangle = width × height*

Algebraically, let the width be w cm and the height be h cm. Then the area, A cm^2, is given by

$$A = w \times h$$

Exercise 16.1

(Mainly revision)

Find the areas of the rectangles in questions 1 to 5. The lengths are in metres.

1

18

3

Remember to give your answer in m^2.

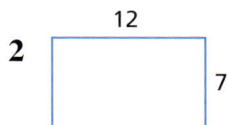

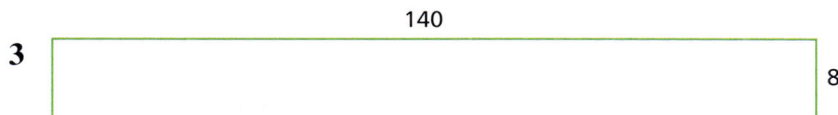

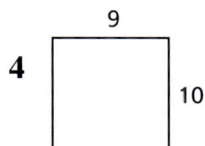

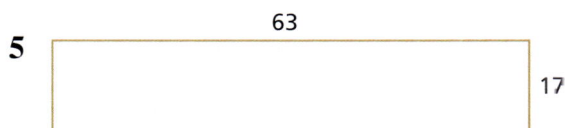

2

12

7

3

140

8

4

9

10

5

63

17

6 A wall is 3 m high and 4 m wide. What is its area?

7 A sheet of A4 size paper is 210 mm wide and 297 mm high. What is its area?

8 A region of land is a rectangle 2 km long and 3 km wide. What is its area?

Fractional areas

Not all rectangles have sides which are a whole number of centimetres! You cannot always fill an area with standard 1 cm^2 tiles! That doesn't matter – if you are tiling, then you can always break the tiles. If you are finding an area, you can multiply together lengths which aren't whole numbers. The same rule for area still holds.

$$A = w \times h$$

Example   A rectangle has width 4 cm and height $2\frac{1}{2}$ cm. Find the area.

To find the area, use the formula $A = w \times h$. Multiply width and height together.

$$4 \times 2\tfrac{1}{2} = 10$$

So the area is 10 cm².

If you look at the diagram, you can see that the rectangle contains 8 whole tiles and 4 half tiles.

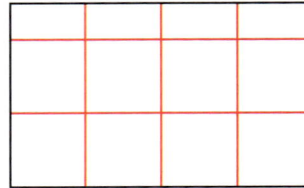

$$8 + 4 \times \tfrac{1}{2} = 10$$

Both lengths could be fractional, as in the next example.

Example   This rectangle has width 4.4 cm and height 2.7 cm. Find the area.

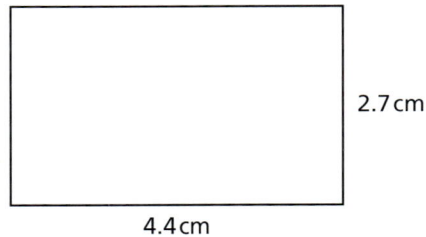

2.7 cm

4.4 cm

Multiply width and height.

$$4.4 \times 2.7 = 11.88$$

So the area is 11.88 cm².

Exercise 16.2

Find the areas of the rectangles in questions 1 to 5. The lengths are in centimetres.

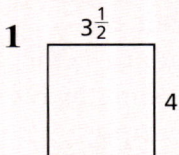

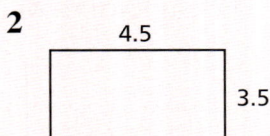

1 $3\frac{1}{2}$

4

2   4.5

3.5

3

1.8

2.9

4

4.8

3.6

5

$2\frac{1}{2}$

$3\frac{1}{2}$

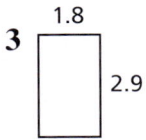

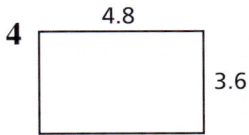

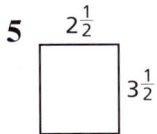

6 A table top is a rectangle which is 20.5 cm wide and 30.5 cm long. What is its area?
7 The case of a computer disk is a rectangle 8.8 cm by 9.3 cm. What is its area?
8 The floor of a room is a rectangle which is 21.4 feet long and 15.5 feet wide. What is its area?

Different units

Make sure that both height and width are in the same units – for example both in centimetres, or both in metres.

Example The rectangle shown is 45 cm by 1.1 m. Find its area.

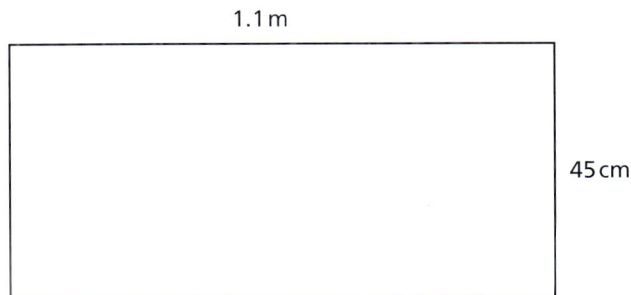

1.1 m

45 cm

Converting to centimetres, 1.1 m is equal to 110 cm.
Use the formula $A = w \times h$.

$$45 \times 110 = 4950$$

So the area is 4950 cm².
Converting to metres, 45 cm is equal to 0.45 m.

$$0.45 \times 1.1 = 0.495$$

So the area is 0.495 m².

Exercise 16.3

Find the areas of the rectangles in questions 1 to 5. Write the answer in the unit given.

1 in cm²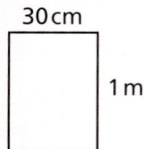

2 in m²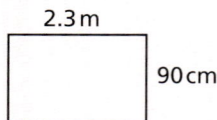

3 in m²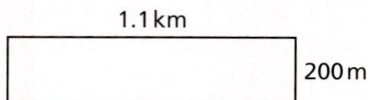

4 in mm²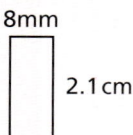

5 in m²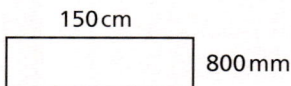

6 A road is 5 m wide. What is the area of a straight stretch of the road which is 3 km long? Give your answer in square metres.

7 A rectangular field is 25 feet wide and 48 yards long. Find its area. Give your answer in square feet. (1 yard = 3 feet)

8 A roll of fax paper is 210 mm wide and 20 m long. What is its area? Give your answer in

    a m²   **b** mm²   **c** cm²

Exercise 16.4   Ma1

Find the area of the walls and ceiling of your bedroom at home. How much paint would you need to cover them?

Parallelograms

Stack a pile of paper on a table. Look at the side facing you. You see a rectangle. Now watch as the pile is slowly pushed over. What has changed, and what is still the same?

- The shape has changed. It is now a parallelogram rather than a rectangle.
- The left- and right-hand sides have changed. The sloping side is now longer than the original vertical side.
- The width of the base is still the same.
- The height of the pile is still the same.
- The amount of paper is still the same. Hence the area of the side is still the same.

We still get the same area, however far we push the paper. The parallelogram has the same area as the original rectangle. So the formula for the area of the parallelogram is the same as that for the rectangle.

- *area of a parallelogram = width × height*

$$A = w \times h$$

> The height must be the *perpendicular* height. Do not multiply the base width by the sloping side. This would give a wrong area; it would be too large!

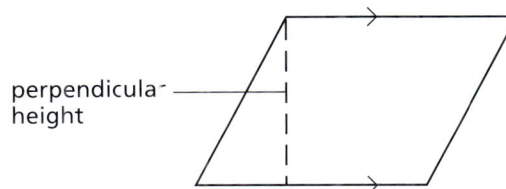

perpendicular height

You can convince yourself of the formula by doing the following exercise.

Exercise 16.5

Cut out a rectangle from a sheet of cardboard. Measure its width and height, and find its area. Make a slanting cut from the bottom left corner to the top side. Shift the triangle from the left hand side to the right. What shape do you have now? What are its base and height? What is its area?

You will need:
- cardboard
- ruler
- scissors

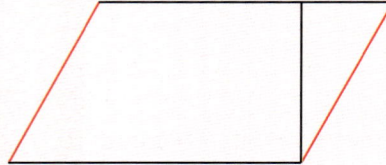

You should now be convinced of the formula for the area of a parallelogram. Use it! Don't go through either of the procedures above, just use the formula.

Notice that this is the first non-rectangular shape whose area you have found.

Example A parallelogram has width 4 cm and perpendicular height 3.5 cm. Find its area.

Note again that the *perpendicular* height is 3.5 cm.
Use the formula, $A = w \times h$.

$$4 \times 3.5 = 14$$

The area is 14 cm².

Exercise 16.6

Find the areas of the parallelograms of questions 1 to 5. The lengths are in centimetres.

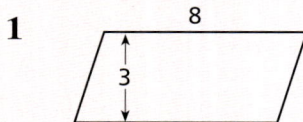

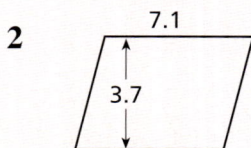

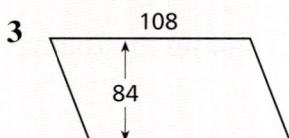

1 8, 3

2 7.1, 3.7

3 108, 84

4

22 63

5

49 **31**

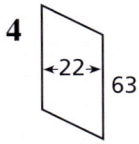

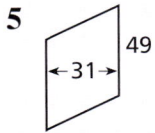

6 A rhombus (which is a special sort of parallelogram) has base 7 cm and perpendicular height 4 cm. What is its area?

7 Michelle is making herself a pair of earrings in the shape of a rhombus. She makes slanting cuts in a rectangular strip of plastic, as shown. What is the area of each earring?

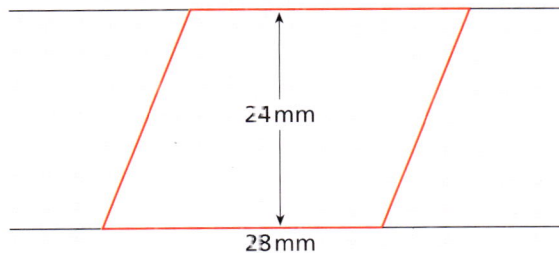

24 mm

28 mm

8 A parallelogram has base 34 mm and perpendicular height 1.3 cm. Find its area, giving your answer in square millimetres.

9 A parallelogram has base 1.3 km and perpendicular height 420 m. Find its area, giving your answer
  a in km²   **b** in m²

Triangles

Perpendicular height is also known as 'altitude'.

Look at this triangle, with base b and perpendicular height h. How can we find its area?

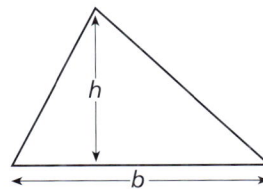

h

b

You will need:
- card
- ruler
- pencil
- scissors

    Make two copies of the triangle, and cut them out. Swivel one triangle round and join it to the other triangle as shown.
    What do you have now?

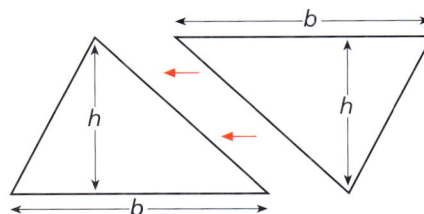

h

b

b

h

You should have a parallelogram, also with base *b* and height *h*. We know that the area of the parallelogram is the product of the base and the height, that is, $b \times h$, or bh.

The parallelogram is made out of two triangles, so the area of one triangle is half the area of the parallelogram: $\frac{1}{2}bh$.

● *area of a triangle $= \frac{1}{2} \times$ base $\times$ height*
$$A = \tfrac{1}{2}bh$$

Note that the height must be the perpendicular height of the triangle.

Exercise 16.7

This exercise is another way to find the formula for the area of a triangle. Draw a triangle *ABC*, and find the midpoints *D* and *E* of *AB* and *AC*. Cut out the triangle, and then cut along *DE*. Arrange the pieces so that *AE* lies along *EC*, as shown.

You will need:

• card
• ruler
• pencil
• scissors

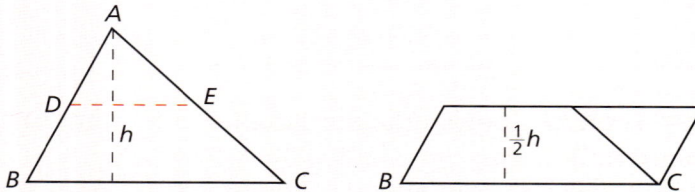

What shape do you have now? It should be a parallelogram. The area of this parallelogram is the same as the area of the triangle. The base of the parallelogram is the base of the original triangle. The height of the parallelogram is half the height of the triangle. Now use the formula for the area of a parallelogram.

area = base $\times$ height of parallelogram
= base $\times \frac{1}{2} \times$ height of triangle

You should be convinced of the formula now!

Example   This triangle has base 10 m and perpendicular height 9 m. What is its area?

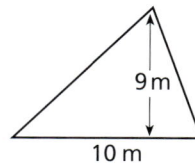

Use the formula, $A = \frac{1}{2}bh$.

$$\tfrac{1}{2} \times 10 \times 9 = 45$$

So the area is 45 m².

Exercise 16.8

Find the areas of the triangles in questions 1 to 5. The lengths are in metres.

1
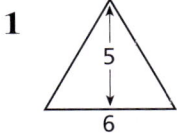

2
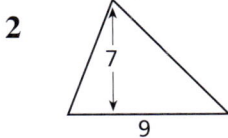

3
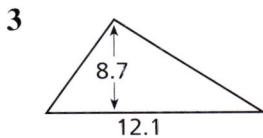

4
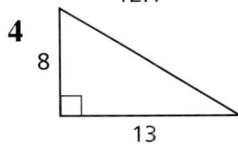

5
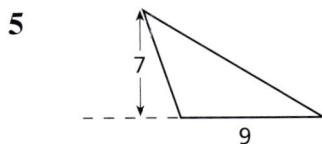

6 The perpendicular height of an equilateral triangle is approximately 0.87 times its side. Find the approximate area of an equilateral triangle with side 8 cm.

7 A triangle has base 43 cm and perpendicular height 1.3 m. Find its area, giving your answer in square centimetres.

8 A triangle has base 46 mm and height 2.7 cm. Find its area, giving your answer in square centimetres.

9 A right-angled triangle has sides 3 cm, 4 cm and 5 cm. The right-angle is between the sides of length 3 cm and 4 cm. What is the area of the triangle?

10 A right-angled triangle has sides of length 5 cm, 12 cm and 13 cm. What is its area?

Finding lengths

With our rectangles, parallelograms and triangles, we have been given two lengths and found an area. We can reverse this – given one length and the area, we can find the other length.

In the three formulae we have, two lengths are multiplied to give an area. What is the opposite of multiplication? It is division, so we *divide* an area by a length to find another length.

Examples   A rectangle has width 12 cm, and area 72 cm². What is the height?

To find the area, we multiply the lengths. So, to find the height, we *divide* the area by the width.

$$\text{area} = \text{width} \times \text{height}$$

So

$$\text{height} = \text{area} \div \text{width}$$
$$72 \div 12 = 6$$

The height is 6 cm.

....

A parallelogram has area 24 cm² and height 8 cm. What is the width?

Here we shall be more formal. Let the width be x cm. The product of the width and the height is the area. This gives us an equation, like ones you have already solved.

$$8 \times x = 24$$

Divide both sides by 8, obtaining $x = 3$.
   The width is 3 cm.

....

A triangle has area 7 cm² and base 4 cm. What is its height?

This one is trickier, as the formula for the area of a triangle involves $\frac{1}{2}$.

$$\text{area} = \tfrac{1}{2} \times \text{base} \times \text{height}$$

To reverse this formula, first multiply both sides by 2.

$$2 \times \text{area} = \text{base} \times \text{height}$$

Then divide by the base.

$$\text{height} = \frac{2 \times \text{area}}{\text{base}}$$

$$\frac{2 \times 7}{4} = \frac{14}{4} = 3.5$$

So the height is 3.5 cm
   We can check the answer by putting it back into the formula for the area of a triangle.

$$\text{area} = \tfrac{1}{2}\text{base} \times \text{height} = \tfrac{1}{2} \times 3.5 \,\text{cm} \times 4 \,\text{cm} = 7 \,\text{cm}^2$$

So the answer is correct.

Remember:
$5 \times x = 35$
$x = 35 \div 5$
$x = 7$

Exercise 16.9

1 A parallelogram has area $120\,m^2$ and base $8\,m$. What is its height?
2 A rectangular field has one side $150\,m$. If its area is $9000\,m^2$ find the other side.
3 A triangle has area $5\,m^2$ and a base of $2\,m$. What is its height?
4 A triangle has area $10\,cm^2$ and height $4\,cm$. What is its base?

Copy the table below, which gives the lengths and areas for various shapes. Some numbers are missing. Work them out and complete the table.

	shape	base (cm)	height (cm)	area (cm²)
5	rectangle	18		414
6	triangle	5		35
7	parallelogram		12	108
8	rectangle		15	54
9	triangle	9.3		31.62
10	parallelogram		2.1	0.42

Combinations of shapes

Some more complicated shapes can be built up from basic rectangles, parallelograms or triangles. To find the area of these shapes, find the areas of the basic shapes and add them together.

Example Find the area of this shape.

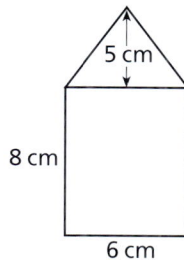

The shape is a triangle on top of a rectangle.

$$\text{area of triangle} = \tfrac{1}{2} \times 6\,cm \times 5\,cm = 15\,cm^2$$
$$\text{area of rectangle} = 6\,cm \times 8\,cm = 48\,cm^2$$
$$15 + 48 = 63$$

So the area of the shape is $63\,cm^2$.

Exercise 16.10

Find the areas of the shapes in questions 1 to 5. The lengths are in centimetres.

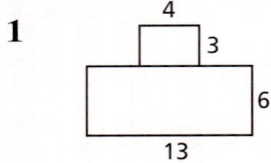

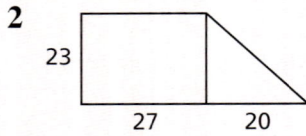

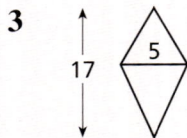

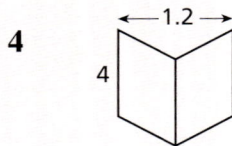

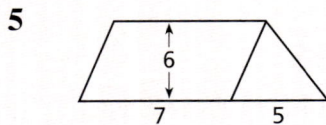

1

```
   4
  ___
 |   | 3
_|___|___
|        | 6
|_____|
    13
```

2

```
 _____
|          /|
23|        / |
|_____/__|
   27      20
```

3

```
  ^
  |  /\
17|  5\/
  |  /\
  |  \/
  v
```

4

```
 <-1.2->
  \    /
4  \  /
    \/
```

5

```
   ____
  /  ^ \
 /   6  \
/____v___\
  7     5
```

6 The dots on the diagram below are 1 cm apart. Find the areas of the shapes.

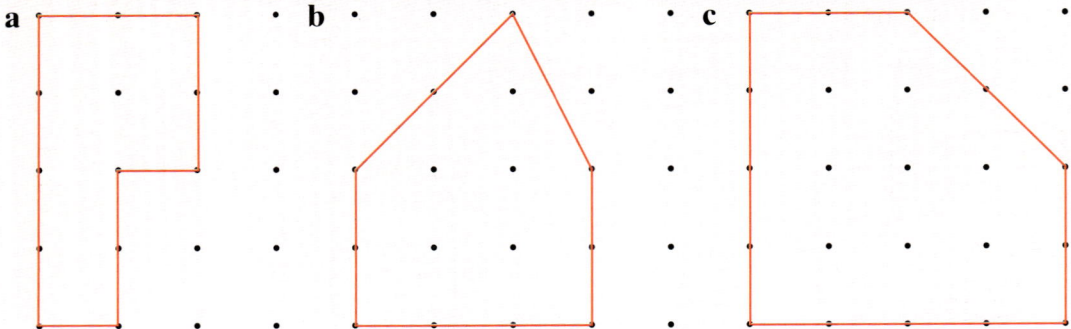

a **b** **c**

7 The diagonals of a rhombus are of length 4 cm and 9 cm. Find the area of the rhombus. (The diagonals of a rhombus meet at 90°.)

8 A regular pentagon is made up of five isosceles triangles which have base 4 cm and approximate height 2.75 cm. Find the approximate area of the pentagon.

9 Bert's garden has a lawn which is a rectangle 14 m by 8 m. The lawn is surrounded by a path which is 1 m wide. What is the area of lawn and path?

10 Maria has a photograph which she wants to frame. The photograph is a rectangle 12 cm by 8 cm, and she wants it to be mounted so that there is a 2 cm gap between the photo and the frame. What is the area of the photo and mount?

In some cases it is quicker to *subtract* one basic area from another.

Example Find the area of this pink shape.

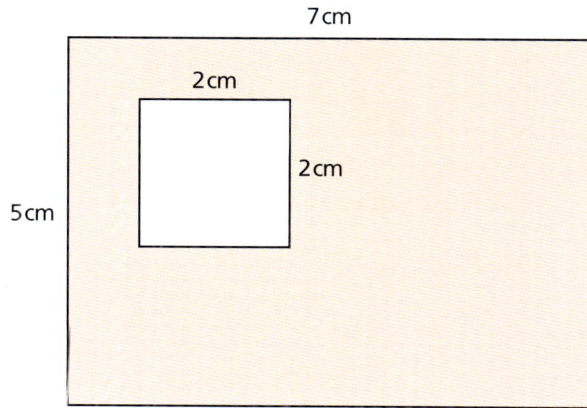

7 cm

2 cm

2 cm

5 cm

We could cut the shape up into four rectangles. But it is easier to think of it as a rectangle with a square removed from the inside.

$$\text{area of rectangle} = 5\,\text{cm} \times 7\,\text{cm} = 35\,\text{cm}^2$$
$$\text{area of square} = 2\,\text{cm} \times 2\,\text{cm} = 4\,\text{cm}^2$$
$$35 - 4 = 31$$

So the area of the shape is 31 cm².

Exercise 16.11

In each of questions 1 to 5 find the area of the shaded shape. The lengths are in centimetres.

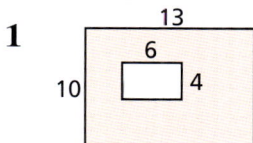

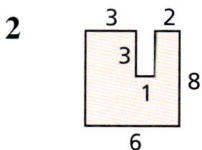

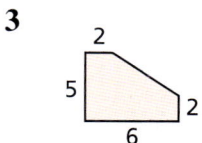

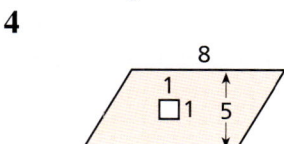

1

13

6

10

4

2

3   2

3

8

1

6

3

2

5

2

6

4

8

1

1   5

5

8

←3→ 6

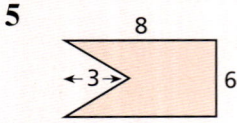

6 My garden is a rectangle which is 15 m by 30 m. There are flower beds round the edges, and there is a lawn in the middle which is a rectangle measuring 14 m by 29 m. What is the total area of the flower beds?

7 Find the area of the L-shape shown, in two ways. Check that your answers are the same.

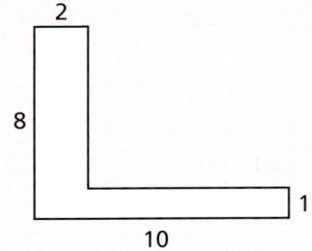

2

8

1

10

a By finding the area of two rectangles and adding
b By subtracting the area of one rectangle from another

8 Lizzie wants to make a metal tray. She takes a square of metal of side 40 cm. From each corner she cuts off a square of side 2 cm.

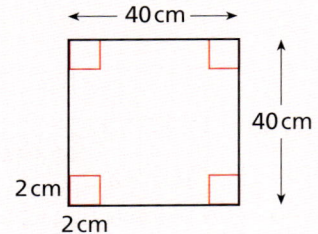

←— 40 cm —→

40 cm

2 cm

2 cm

a What is the remaining area?
b The flaps are now folded up to make the sides of the tray. What is the area of the base of the tray?

Trapezia

A trapezium is a quadrilateral with one pair of parallel sides. We can find its area by rearranging it into a parallelogram.

Say the perpendicular height is h and the parallel sides have lengths a and b. Cut the trapezium in two, by a line parallel to the parallel sides, and half-way between them. Rearrange the parts as shown. We now have a parallelogram.

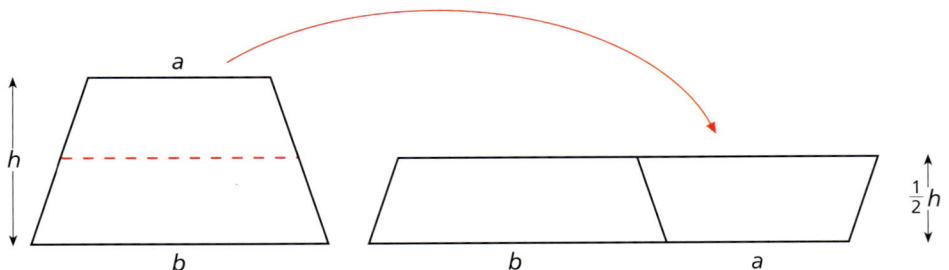

The plural of trapezium is trapezia.

a

h

b

b

a

$\frac{1}{2}h$

Remember:

To do the addition first, you put it in brackets.
(See chapter 5.)

base of parallelogram $= a + b$
height of parallelogram $= \frac{1}{2}h$

So the area of the parallelogram, and hence of the trapezium, is $\frac{1}{2}h(a + b)$. Note that you must add the parallel sides first, and then multiply by $\frac{1}{2}h$.

Exercise 16.12

This is another way to find a formula for the area of a trapezium. Draw a trapezium $ABCD$, with height h and parallel sides $AB = a$ and $CD = b$.

You will need:
• card
• ruler
• pencil
• scissors

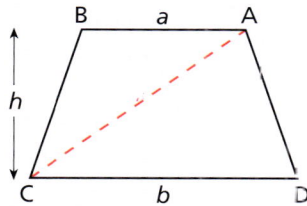

Cut the trapezium along AC. You now have two triangles.

1 What is the area of triangle ABC?
2 What is the area of triangle ADC?
3 Add these formulae. You should get an expression which you can show is equal to $\frac{1}{2}h(a + b)$.

Example Find the area of the trapezium shown.

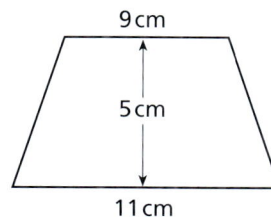

The parallel sides are 9 cm and 11 cm. Add these to obtain 20 cm. Now multiply by $\frac{1}{2}h$.

$$\tfrac{1}{2} \times 5 \times 20 = 50$$

So the area is 50 cm².

Exercise 16.13

Find the area of the trapezia in questions 1 to 5. Lengths are in millimetres.

1

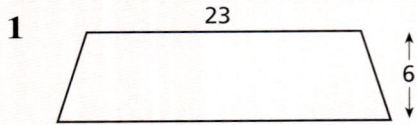

2

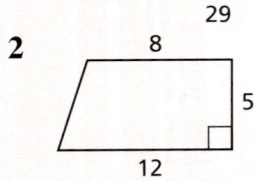

3

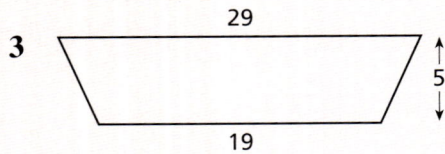

4

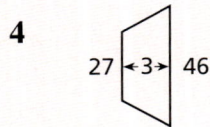

5

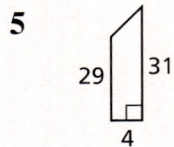

6 A swimming pool is 50 m long. The shallow end is 1 m deep, and it slopes steadily to the deep end, which is 3 m deep. Find the area of a side wall.

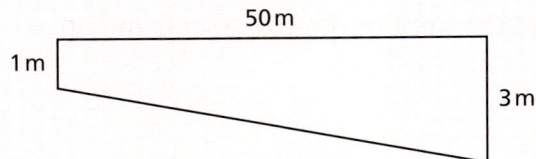

7 The side wall of a hut is shown below. The front wall is $2\frac{1}{2}$ m high, the back wall is 2 m high and the distance between the walls is 2 m. Find the area of the side wall.

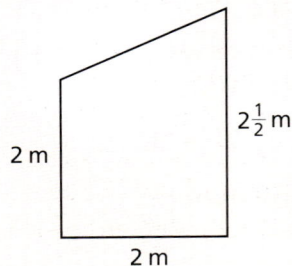

8 The diagram below shows a side window of a car. Find its area.

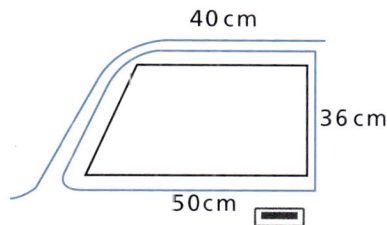

9 The two parallel sides of a trapezium have lengths 9 cm and 21 cm. The area of the trapezium is 120 cm². What is the height of the trapezium?

10 A trapezium has height 4 m and area 10 m². One of the parallel sides is 2 m. What is the other?

SUMMARY

- To find the area of a rectangle multiply its two sides together. The lengths may be fractional. Be careful to measure both lengths in the same units.
- To find the area of a parallelogram multiply its base by its height. Make sure that you take the perpendicular height, not the length of a slanting side.
- To find the area of a triangle multiply its base by its height, then halve.
- If you know the area of a rectangle, parallelogram or triangle, and one of the lengths, then you can find the other length.
- The area of a compound shape can be found by cutting it into simpler shapes, and then adding their areas. Sometimes it is easier to subtract one area from another.
- To find the area of a trapezium, add the parallel sides, then multiply by the height and by $\frac{1}{2}$.

Exercise 16A

1 A rectangle is 15.3 cm by 17.4 cm. Find its area.
2 A long roll of carpet is 2.5 feet wide and 50 yards long. What is its area? Give your answer in square feet. (1 yard = 3 feet)
3 A parallelogram has perpendicular height $3\frac{1}{2}$ m and base $7\frac{1}{4}$ m. Find its area.
4 A triangle has perpendicular height 4 m and base 7 m. Find its area.
5 The area of a rectangle is 120 m² and one side is 24 m. Find the other side.
6 A triangle has height 5.2 m and area 4.42 m². Find its base.

7 Find the area of the shape shown below. Lengths are in metres.

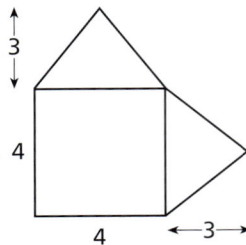

8 Find the area of the shape shown below. Lengths are in centimetres.

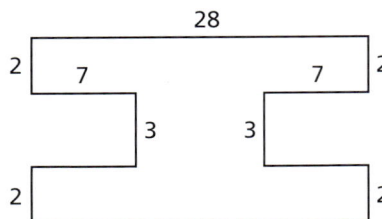

9 A regular hexagon is made from six isosceles triangles which have base 12 cm and approximate perpendicular height 10.4 cm. Find its approximate area.

10 The parallel sides of a trapezium are 5 cm and 17 cm, and its height is 7 cm. Find its area.

Exercise 16B

1 The case of a computer floppy disk is a rectangle, $3\frac{1}{2}$ inches by $3\frac{2}{3}$ inches. Find its area.

2 Find the area of a rectangle which is 3 mm by 22 cm. Give your answer in square centimetres.

3 Find the area of a parallelogram which is 26 cm high and has base 14 cm.

4 A triangle has height 22.3 cm and base 47 cm. Find its area.

5 A rectangular field has area 38 000 m². One side is 190 m. Find the other side.

6 The area of a triangle is 36 cm², and its base is 48 mm. Find its height in millimetres.

7 Find the area of the shape shown below. Lengths are in centimetres.

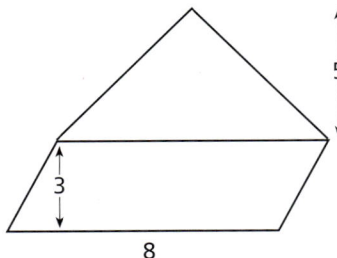

8 The shape shown below represents a castle wall. Lengths are in metres. Find its area.

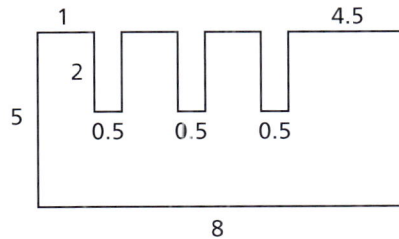

9 The parallel sides of a trapezium are 3 m and 6 m, and its height is 8 m. Find its area.

10 The parallel sides of a trapezium are 15 cm and 17 cm, and its area is 45 cm². Find its height.

Exercise 16C   Ma 1

The diagram below shows a kite. You don't know the lengths of the sides, but the two diagonals have lengths 10 cm and 16 cm.

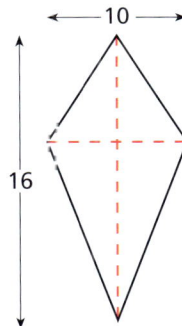

1 What is the area of the kite?

2 Suppose a kite has diagonals of lengths a and b. Can you find a formula for the area of the kite?

Exercise 16D Ma1

A farmer has 100 m of fencing with which to make a sheep pen. He can use part of a long stone wall as one side of a rectangle, and arrange the fencing for the other three sides, as shown below. Obviously he wants to enclose as large an area as possible. Can you advise him?

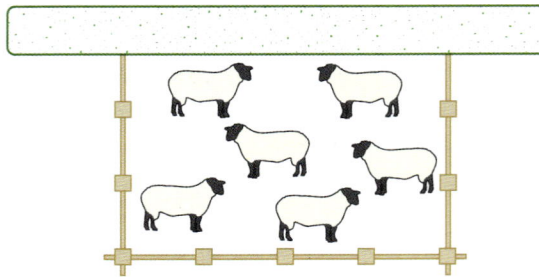

Two of the sides are perpendicular to the wall. Try different lengths for these sides, find the length of the side parallel to the wall, and hence find the area. You could use the table below.

perpendicular sides (m)	parallel side (m)	area (m²)
10	80	800
20		

17 Equations

A group of eight boys are having a 'tug of war'. Michael, Brian, Dario and David are on one team. Edward, Farzook, George and Henry are on the other. We know that the first four can pull with a combined force of 2000 newtons, and that Edward and Farzook have a combined force of 1200 newtons. George and Henry are newcomers to the sport, but they are identical twins, and so we assume they pull with identical force.

When they compete, they find that the two teams are exactly balanced. What is the force of the identical twins?

There is something we don't know, the force of each twin. This is an **unknown**. We write down the facts about this unknown, and we obtain an **equation**. This chapter is about setting up and solving equations.

In this case, let each twin pull with a force of x newtons, then the following equation must be true.

$$1200 + 2x = 2000$$

The solution is $x = 400$. Each twin has a force of 400 newtons.

You have solved equations like this one, and like the one below.

$$3x + 5 = 26$$

How did you do it?

First you subtracted 5 from both sides.

$$3x = 26 - 5$$
$$= 21$$

Then you divided by 3.

$$x = 21 \div 3$$
$$= 7$$

This is the answer. To check that it is correct, substitute 7 for x in the left-hand side of the original expression.

$$3 \times 7 + 5 = 21 + 5 = 26$$

So the answer $x = 7$ is correct.

> The = sign was first used by an English mathematician, Robert Recorde, in a book called *The Whetstone of Witte*, published in 1557. The sign represents two parallel lines, 'bicause noe 2. thynges, can be moare equalle'.

One way to look at the solution is by a **function machine**. The machine for the left-hand side takes x, multiplies it by 3, then adds 5. The result is 26.

$$x \rightarrow \text{multiply by 3} \rightarrow \text{add 5} \rightarrow 26$$

To find the value of x we go backwards. Start with 26. The opposite to 'add 5' is 'subtract 5'. The opposite to 'multiply by 3' is 'divide by 3'.

$$7 \leftarrow \text{divide by 3} \leftarrow \text{subtract 5} \leftarrow 26$$

When we subtract 5 from 26 we get 21. When we divide by 3 we get the answer, 7.

Exercise 17.1

(Mainly revision)
Solve these equations. Check your answers.

1 $2x + 5 = 17$ **4** $2x - 7 = 11$ **7** $3x + 15 = 27$ **9** $8m - 5 = 59$
2 $3x + 2 = 29$ **5** $3a - 5 = 10$ **8** $5y + 14 = 49$ **10** $11a - 24 = 97$
3 $5y - 7 = 13$ **6** $4x - 5 = 11$

Equations with brackets

Remember chapter 5.

We have already looked at expressions involving brackets. Here we shall see how we can solve equations with brackets. Look at this equation.

$$3(x + 8) = 30$$

Because of the brackets, the addition is done before the multiplication. The function machine for the left-hand side takes x, adds 8, then multiplies by 3. The result is 30.

$$x \rightarrow \text{add } 8 \rightarrow \text{multiply by } 3 \rightarrow 30$$

Reverse the direction of this function machine.

$$2 \leftarrow \text{subtract } 8 \leftarrow \text{divide by } 3 \leftarrow 30$$

Following these instructions, divide 30 by 3, obtaining 10. Subtract 8, obtaining 2. This is the solution of the equation.

Example   Solve this equation without using function machines. Check your answer.

$$2(x - 7) = 18$$

Divide both sides by 2.

$$x - 7 = 18 \div 2$$
$$= 9$$

We no longer need brackets, as there is only one operation.

Add 7.

$$x = 9 + 7$$
$$= 16$$

Check: substitute 16 for x in the left-hand side of the original equation.

$$2(16 - 7) = 2 \times 9$$
$$= 18$$

So the answer is correct.

Note. There is another way to solve this sort of equation. In chapter 5 we learnt how to multiply out brackets. Look again at the equation.

$$2(x - 7) = 18$$

Expand the brackets on the left-hand side.

$$2x - 14 = 18$$

Add 14 to both sides.

$$2x = 32$$

Divide both sides by 2.

$$x = 16$$

The answers are the same. Use whichever method you prefer.

Exercise 17.2

Solve these equations. Check your answers.

1 $3(x + 8) = 36$ **4** $2(x - 3) = 14$ **7** $3(5y + 3) = 39$ **9** $2(2p + 5) = 22$
2 $2(x + 4) = 26$ **5** $3(x - 6) = 0$ **8** $5(2z - 3) = 35$ **10** $3(2q - 6) = 0$
3 $5(v - 2) = 10$ **6** $2(3x + 4) = 32$

If there is more than one pair of brackets in an equation, you have to solve it by expanding the brackets first.

Example   Solve this equation. Check your answer.

$$2(x + 3) + 3(x + 5) = 31$$

Open out the brackets.

$$2x + 6 + 3x + 15 = 31$$

Collect together like terms.

$$5x + 21 = 31$$
$$5x = 10 \quad \text{(subtracting 21)}$$
$$x = 2 \quad \text{(dividing by 5)}$$

Remember chapter 5.

Check: put $x = 2$ in the left-hand side of the original equation.

$$2(2 + 3) + 3(2 + 5) = 2\times5 + 3\times7$$
$$= 10 + 21$$
$$= 31 \quad \text{Correct.}$$

Exercise 17.3

Solve these equations. Check your answers.

1 $2(x + 3) + 4(x + 7) = 52$ **6** $2(3y + 2) + 3(2y + 1) = 31$
2 $3(y + 7) + 4(y + 3) = 54$ **7** $3(3x + 5) + 4(2x + 3) = 112$
3 $2(x - 3) + 3(x + 7) = 35$ **8** $4(y + 5) + y = 75$
4 $5(z - 2) + 3(z + 5) = 37$ **9** $2(2x - 9) + 4x = 14$
5 $4(p + 2) + 3(p - 7) = 57$ **10** $5(3y + 2) - 7y = 58$

Equations where the unknown appears on both sides

At the beginning of the chapter there were eight boys competing in a tug of war. The game catches on. Eight girls now compete: Anne, Selina, and Cynthia can pull with a combined force of 1400 newtons. Davina can pull with a force of 460 newtons. They are joined by Amy, Frances, Geri and Helene, who are identical *quadruplets*. After a bit of experiment they find that if Amy joins Anne, Selina and Cynthia on one team, and the other three quads join Davina in the other, the two teams will exactly balance.

Can you find an equation to express this? If each of the quads can pull with a force of *y* newtons, can you find an equation in *y*? Can you solve this equation?

One quad, with the first three girls, balances the other three quads with the fourth girl. The total force of the left-hand side is $(1400 + y)$ newtons, and the total force of the right-hand side is $(460 + 3y)$ newtons.

The two sides balance. So the force of the two teams must be equal.

$$1400 + y = 460 + 3y$$

Now, Amy and Frances are two of the identical quads, on opposing teams. The sides will still balance if both of these girls leave simultaneously.

What equation will we have now? On one side are the three girls pulling with a force of 1400 newtons, on the other are Davina with a force of 460 newtons and the two remaining quads with a force of *y* newtons each.

$$1400 = 460 + 2y$$

This is now one of the simpler sorts of equation, in which the unknown *y* only occurs once. Subtract 460 from both sides, then divide by 2.

$$940 = 2y$$
$$470 = y$$

Hence each of the quads can pull with a force of 470 newtons.

In this section we shall solve equations in which the unknown appears twice. They might be like the one we have just done, $1400 + y = 460 + 3y$.

Now we'll do one just by algebra, without bringing in a tug of war.

Example   Solve this equation. Check your answer.

$$5x + 5 = 2x + 23$$

Subtract $2x$ from both sides of this equation (just as we got one girl to leave both teams at the tug of war). When we remove $2x$ from the right-hand side, all that is left is 23. When we remove $2x$ from $5x$, what is left is $3x$.

$$5x + 5 - 2x = 23 \quad \text{(subtracting } 2x \text{ from both sides)}$$
$$3x + 5 = 23 \quad \text{(collecting like terms)}$$

Subtract 5 from both sides, then divide by 3.

$$3x = 18$$
$$x = 6$$

To check your answer, put $x = 6$ into both sides of the original equation. They should be equal!

Left-hand side: $5 \times 6 + 5 = 30 + 5 = 35$
Right-hand side: $2 \times 6 + 23 = 12 + 23 = 35$

So the answer is correct.

> Note that we subtracted the smaller term involving x. If we subtract 5x from both sides, we would get $-3x$. This wouldn't be wrong, it would just make it more complicated.

With equations, there is one rule that must always be obeyed. Whatever we do to one side, we must always do to the other.

● *Do to the left-hand side what you do to the right-hand side.*

Exercise 17.4

Solve these equations. Check your answers.

1 $3x + 2 = 2x + 7$
2 $4y + 5 = 3y + 13$
3 $5t + 1 = 3t + 29$
4 $9v + 4 = 2v + 39$
5 $3p + 5 = p + 31$

6 $2a + 15 = 4a + 3$
7 $m + 19 = 5m + 3$
8 $7n + 13 = 2n + 98$
9 $2c + 100 = 11c + 10$
10 $100x + 1 = 72x + 29$

Negative *x* terms

Let us now look at another tug of war. Ivan, James and Kevin can pull with forces of 450 newtons, 880 newtons and 770 newtons, respectively. They are joined by Laurie, Maurice and Nigel, who are identical *triplets*. Laurie and Maurice join Ivan on one side of the rope, and Nigel joins James and Kevin on the other.

Perhaps Nigel saw that his brothers were totally outclassed by James and Kevin, and so he decided to help his brothers.

The sides balance. But notice that Nigel is pulling in the wrong direction. Instead of helping James and Kevin, he is hindering them. His force should be *subtracted* from theirs, not added.

Let us try to get an equation from this situation. Suppose that each of the three brothers can exert z newtons. Two of the triplets are on one side with Ivan, who has a force of 450 newtons. The total force of that side is $(2z + 450)$ newtons.

On the other side, James and Kevin have a total force of 1650 newtons. Nigel is pulling in the wrong direction, so his effort is subtracted from theirs. The total force on that side is $(1650 - z)$ newtons.

The sides balance, so we have this equation

$$2z + 450 = 1650 - z$$

Now to solve it. Because Nigel is pulling in the wrong direction, he is helping the other side. It would make no difference if he were to move over to the other side, to join his two brothers. The sides would still balance. The equation would become

$$3z + 450 = 1650$$

Now it is one of the simpler equations. Subtract 450 and divide by 3.

$$3z = 1200$$
$$z = 400$$

So each of the triplets pulls with a force of 400 newtons.

The key step in this problem was when we moved the z over to the other side. It was $-z$ on the right-hand side, and $+z$ on the left-hand side. So that when this term changed sides, it changed from negative to positive.

This fits in with our general rule, of *do to the left what you do to the right.* If there is a $-z$ on the right, then to get rid of it we must add z to the right. If we add z to the right, then we must add z to the left also.

The next example doesn't involve tugs of war.

Example Solve this equation. Check your answer.

$$17 - 2x = 3x - 18$$

Let us get the xs together. There is $-2x$ on the left, so to get rid of it we must add $2x$ to the left. If we add $2x$ to the left, we add $2x$ to the right also. On the right we then get a total of $5x$.

$$17 = 3x + 2x - 18$$
$$17 = 5x - 18$$

Now add 18 to both sides, then divide by 5.

$$35 = 5x$$
$$7 = x$$

> Instead of adding $2x$, we could subtract $3x$ from both sides. We would get $-5x$ on the left. This wouldn't be wrong, it just makes things more complicated.

As a check, substitute 7 for x in both sides of the original equation.

Left-hand side: $17 - 2 \times 7 = 17 - 14 = 3$
Right-hand side: $3 \times 7 - 18 = 21 - 18 = 3$

This confirms that our answer is correct.

Exercise 17.5

Solve these equations. Check your answers.

1 $2x + 7 = 40 - x$
2 $4y + 3 = 51 - 2y$
3 $2z + 7 = 27 - 2z$
4 $p + 5 = 7 - p$
5 $10q + 7 = 124 - 3q$
6 $44 - 3a = 3a + 2$
7 $70 - 2b = 9b + 4$
8 $58 - 5c = 7c + 10$
9 $46 - 9m = 6m + 1$
10 $247 - 6n = 8n + 9$

Fractional solutions

The solution to an equation does not have to be a whole number.
It might be a fraction or a decimal. Even very simple equations can
have a fractional solution!

Example Solve this equation. Check your answer.

$$2x = 5$$
$$x = 2\tfrac{1}{2} \quad \text{(dividing by 2)}$$

The solution is not a whole number. But it is still the solution!
  Check: put $x = 2\tfrac{1}{2}$.

$$2 \times 2\tfrac{1}{2} = 5 \quad \text{Correct.}$$

Exercise 17.6

Solve these equations.

1 $2x = 3$

2 $3y = 4$

3 $4z = 1$

4 $5a = 2$

5 $7b = 10$

6 $6c = 7$

7 $4p = 6$

8 $9q = 6$

9 $8r = 10$

10 $2x = \tfrac{1}{2}$

Example Solve this equation.

$$3x + 5 = x + 18$$

Subtract x from both sides, then subtract 5 from both sides.

We subtract the smaller term involving x.

$$2x + 5 = 18$$
$$2x = 13$$

Now divide both sides by 2. The answer is not a whole number.

$$x = 13 \div 2 = 6\tfrac{1}{2} = 6.5$$

Check.

Left-hand side: $3 \times 6.5 + 5 = 19.5 + 5 = 24.5$
Right-hand side: $6.5 + 18 \quad = 24.5$ Correct.

Exercise 17.7

Solve these equations. Check your answers.

1 $3x + 1 = 8$

2 $5y + 3 = 12$

3 $2z - 4 = 9$

4 $4p - 1 = 14$

5 $9q + 3 = 31$

6 $4x + 7 = x + 20$

7 $5y - 3 = 2y + 2$

8 $3z + 4 = 5z - 13$

9 $3p + 8 = 17 - p$

10 $6q - 8 = 13 - 2q$

Using equations to solve problems

You already know that equations can tell a story. These 'stories' are usually called 'problems', and equations are used to solve them. We have had stories about boxes of counters on shelves. In this book we have had stories about tug-of-war teams.

When you are given one of these problems, there is always something you don't know. It might be the number of counters in a box. This is the unknown, and we name it by a letter. You might say: 'let the number of counters in a box be x'.

In the problem, you are given some fact about the unknown. (Otherwise you would have no chance of solving it!) The fact might be: 'three boxes of counters plus 5 extra counters amount to a total of 38 counters'. Put this fact into an equation. For this example it would be

$$3x + 5 = 38$$

Once you have an equation, solve the equation by the techniques you have learnt in this chapter. Then you will have solved the original problem!

Let's look in the same way at an example like the problems in this chapter. Suppose we don't know the pulling force of each of two girls who are identical twins. This is the unknown, and we name it by a letter. You could say: 'let each twin be able to exert a force of x newtons'.

In this problem, the given fact about the unknown is that the tug of war balances and we know some of the forces. The fact might be: 'the twins plus an extra 600 newtons balance a force of 1100 newtons'. Put this fact into an equation.

$$2x + 600 = 1100$$

Then you can solve it.

Now we'll do some different examples.

> The unknown in an equation must always stand for a number.

Examples Harry is £20 richer than Dalbir. Together they have £180. How much does Dalbir have?

The unknown is the amount of money that Dalbir has. Name this x.

Let Dalbir have £x.

Harry has £20 more than this.

Harry has £$(x + 20)$.

The fact we are given is that together they have £180. This gives us our equation.

$$x + (x + 20) = 180$$

Now we can solve this. Remove the brackets, and collect like terms.

$$x + x + 20 = 180$$
$$2x + 20 = 180$$

Subtract 20 and divide by 2.

$$2x = 160$$
$$x = 80$$

This is the answer. Dalbir has £80.

> Remember that x is a number. It is the number of £ that Dalbir has.

Jane is 12 and John is 3. When will Jane be twice as old as John?

The unknown is the number of years before this happens. Let us call this x.

After x years Jane is twice as old as John.

After x years, John will be $x + 3$, and Jane will be $x + 12$.
The fact we are given is that Jane's age will be twice John's age. This gives us our equation. Be careful to use brackets.

$$x + 12 = 2 \times (x + 3)$$

Multiply out the brackets.

$$x + 12 = 2x + 6$$

Subtract x from both sides, then subtract 6 from both sides.

$$12 = x + 6$$
$$6 = x$$

So after 6 years Jane will be twice as old as John.
Check: after 6 years Jane will be 18 and John will be 9. Yes, after this number of years Jane's age will be twice John's.

Exercise 17.8

1 The diagram below shows a triangle, with sides x cm, $(x + 3)$ cm and $(x - 2)$ cm. The perimeter of the triangle is 19 cm. Find x.

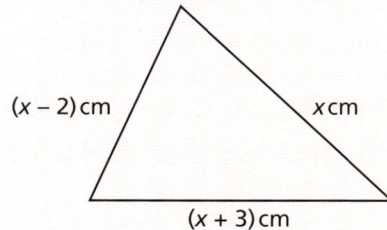

(x – 2) cm   x cm
(x + 3) cm

> Remember that the sum of the angles of a triangle is 180°.

2 The angles of a triangle are $2x$, $(x + 10°)$ and $(2x - 20°)$. Find angle x.

3 The angles round a point are $50°$, x, $(3x + 40°)$ and $(4x + 30°)$. Find angle x.

4 The diagram below shows an equilateral triangle and a regular pentagon, each with side x cm. The perimeter of the pentagon is 40 cm greater than the perimeter of the triangle. Find x.

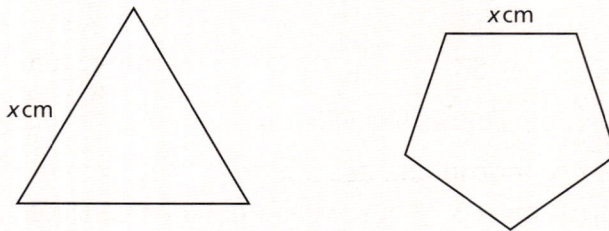

x cm   x cm

> Remember that the perimeter of a shape is the length around it.

5 The diagram below shows a triangle and a square, each with side y cm. The sum of their perimeters is 560 cm. Find y.

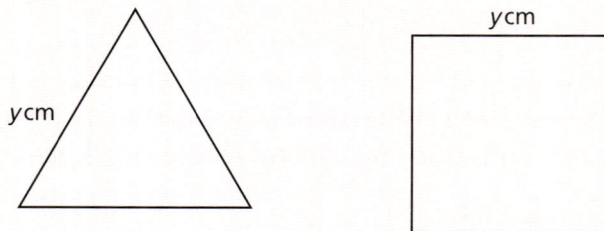

y cm   y cm

6 Delia has twice as many grapes as Alice. They have 129 grapes between them. How many has Alice?

7 Jasper is 5 years younger than Simone. The sum of their ages is 35. How old is Jasper?

8 I buy several 25p and 26p stamps at a post office. I buy six more 25p stamps than 26p stamps. The total cost is 864p. How many of each sort have I bought?

9 A cake costs 5p more than a scone. I buy eight cakes and seven scones for £4.90. What is the cost of a scone?

10 Two children play a number game. The first tells the second to: 'think of a number, double it, add 7'. If the final number is 19, what was the original number?

11 With the money I have at the moment. I could either buy five cakes and have 17p change, or I could buy three cakes and have 97p change. What is the cost of a cake?

12 Seven cars weigh 2400 kg more than four cars. How much does each car weigh?

13 A brother is 5 years older than his sister. In two years time he will be twice as old as her. How old is the sister now?

14 Manuel has three times as much money as Norman. If Manuel gives £5 to Norman, then he will have only twice as much money. How much does Norman have now?

15 Jane is 2 years older than Abe, and Abe is 3 years older than Ngaire. The sum of their ages is 68. How old is Ngaire?

SUMMARY

■ To solve an equation, reverse the steps by which the equation was formed.
■ Be sure to do the same operation to both sides of the equation.
■ If the **unknown** occurs on both sides of the equation, get rid of one occurrence. To remove a positive occurrence of the unknown, *subtract* it. To remove a negative occurrence, *add* it. Of course, you must do the same thing to the other side.
■ The answer to an equation might be a fraction.
■ When writing an equation to solve a problem, use a letter to represent the unknown. The letter should stand for a number, for example x kg, or y years.

Exercise 17A

Solve the equations in questions 1 to 8. Check your answers.

1 $2x + 7 = 23$
2 $4y - 8 = 32$
3 $3(z - 5) = 18$
4 $4(2x + 3) + 5(3x + 1) = 86$
5 $6x = 15$
6 $2y + 13 = 4y + 5$
7 $19 - 2z = 3z + 4$
8 $3z + 12 = 6z + 2$
9 A computer costs £200 more than a printer. Together they cost £1700. How much does the printer cost?
10 Ten years ago a man was three times as old as his son. He is now twice as old. Find their present ages.

Exercise 17B

Solve the equations in question 1 to 8. Check your answers.

1 $5p - 3 = 32$
2 $3q + 4 = 103$
3 $2(3r + 7) = 54$
4 $5(2m - 8) = 3(3m + 9)$
5 $4n = 11$
6 $3p - 25 = p + 7$
7 $1 + 3q = 22 - 4q$
8 $8r - 14 = 30 - 3r$
9 In a certain week, Kevin cycles twice as far as he walks. If he travels a total distance of 21 miles, how far has he walked?
10 Ten years ago, twice Anita's age would have been the same as it will be in ten years time. How old is she now?

Exercise 17C Ma 1

In the last section of this chapter you found the equations to fit the stories. Now go in the other direction! Find stories to fit the equation! You might build upon the tug-of-war stories.

$3x + 1000 = 3480$ (this could involve identical triplets)
$x + 3440 = 4x + 410$ (this could involve identical quintuplets)
$288 - x = 44 + 3x$
$x + 28 = 2(x + 8)$ (this could involve ages of a mother and son)

Exercise 17D Ma 1

al-Khwarizmi, an Arab mathematician, wrote a book called

Kitab al-jabr wa al-muqabalah

Say *al-jabr* quickly. What word do we get from it?
   The title means *The book of shifting and balancing*. It is about solving equations. 'Shifting' (*al-jabr*) means moving terms from one side of the equation to the other. 'Balancing' (*al-muqabalah*) means simplifying the terms on one side of the equation.

1 Look at the solution of the following equation. Which steps are *al-jabr* and which are *al-muqabalah*?

$3x + 4 = 37$
$3x = 33$
$x = 11.$

2 Write out the solution to the equation $5x - 3 = 62$, saying which steps are *al-jabr* and which are *al-muqabalah*.
3 Write out the solution to the equation $3x + 5 = x + 37$, saying which steps are *al-jabr* and which are *al-muqabalah*.
4 Write down an equation of your own, and solve it by *al-jabr* and *al-muqabalah*.

al-Khwarizmi lived in Baghdad around AD 825. Try saying his name quickly! It became *algorithm*, which means a process for solving a problem.

18 Circles

Lots of common phrases involve words about circles.

> 'Don't fly off at a tangent!'
> The private sector of the economy.
> 'We all live within a 10 mile radius of the school.'

Parts of a circle

When you draw a circle with compasses, the pencil is a fixed distance from the point of the compasses. The path traced by the pencil is called a circle. All the points on a circle are at a fixed distance from a fixed point. Here are the official words.

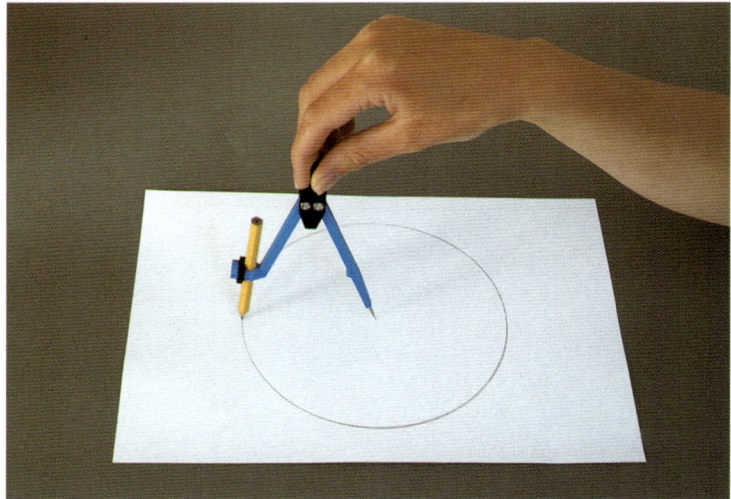

- *The **centre** is the fixed point. It is in the middle of the circle.*
- *The **radius** is the fixed distance. It is the length of a line from the centre to a point on the circle. The word radius has two uses: it means the line itself, and also the length of the line.*

The plural of radius is radii.

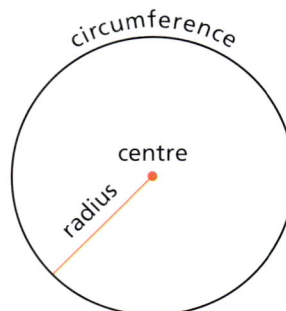

- *The **circumference** is the distance around the circle.*
- *You use **compasses** to draw circles.*

Exercise 18.1

A goat is on the end of a rope, whose other end is fixed to a point on a field. The goat walks round and round, keeping the rope taut. What are the mathematical names for the following?

a The path that the goat follows

b Where the other end of the rope is fixed

c The length of the rope

Now take two points on the circle, *A* and *B*. There are some more words for you to know!

● *The **chord** AB is the straight line joining A and B.*

● *The **arc** AB is the part of the circle between* A *and* B.

> Note that there are two arcs of the circle between A and B. Normally we take the shorter arc.

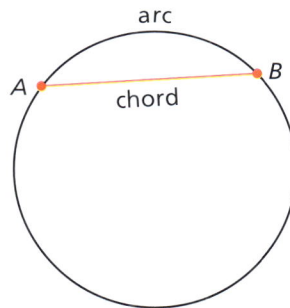

Suppose now that *A* and *B* are on the circle, on opposite sides of the centre. The length of *AB* is at its greatest.

● *A **diameter** is a line across the circle through the centre. The length of the diameter is twice the radius.*

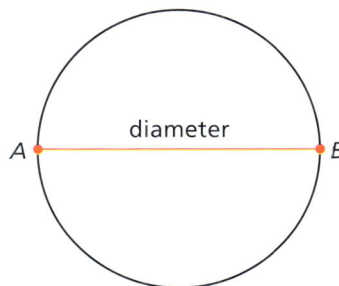

> So a diameter is a special sort of chord.

A chord is a line which meets the circle twice. A line could meet the circle just once. There is a special name for this.

● *A **tangent** is a line which just touches a circle.*

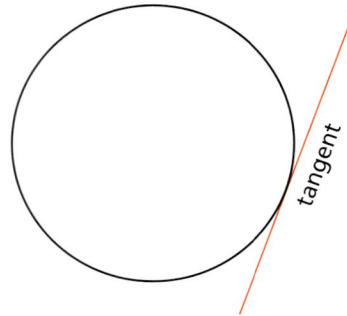

tangent

Exercise 18.2

1 The French phrase *arc-en-ciel* means 'arc-in-the-sky'. What does it refer to?

2 Suppose you are swinging a conker round on the end of a string. What path does it follow? The string suddenly breaks. What path does the conker immediately follow?

3 You are cycling along a straight line on a flat road. What is the relationship between the line and your front wheel?

4 This picture shows Robin Hood. When stretched with an arrow, ready to shoot, the wood of his bow forms part of a circle. What is the mathematical name for this part? What is the mathematical name for the unstretched string?

5 A plank is balanced on circular rollers. What is the mathematical name for the distance of the plank above the ground?

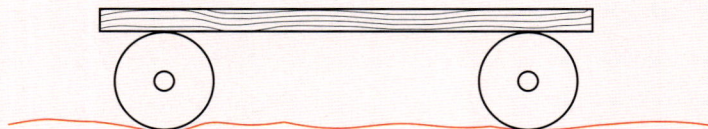

So far we have given names for lines connected with circles. There are also special regions connected with circles.

● *A **sector** is the region between two radii.*

So a semi-circle is a segment cut off by a diameter.

- *A **segment** is the region between an arc and a chord.*
- *A **semi-circle** is half a circle.*

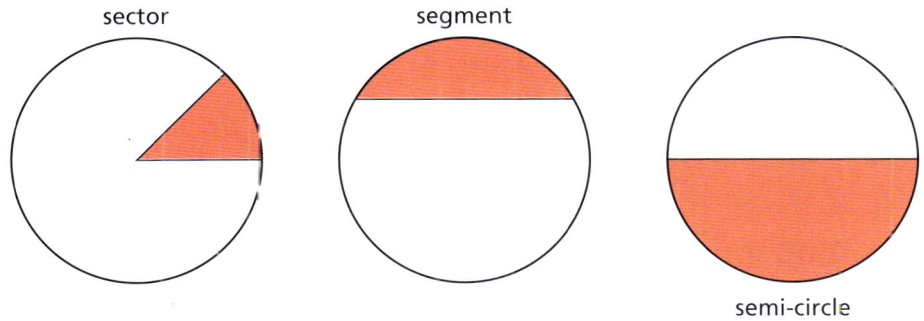

sector   segment

semi-circle

Exercise 18.3

For each of the diagrams in questions 1 to 5 give the name for the part of the circle shown in red.

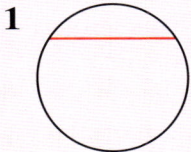

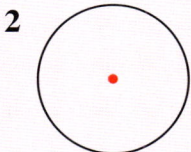

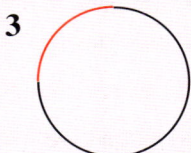

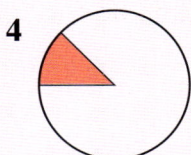

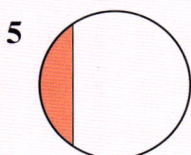

1

2

3

4

5

6 In a café you buy a slice of a circular cake. What word describes the shape of the top?

7 You can buy what are called grapefruit segments. The picture shows what you might see when you open the tin. The shapes are not segments. What should they be called?

8 What shape is your protractor?

9 What shapes is a dartboard divided into?

10 What region of a circle is both a sector and a segment?

Length of the circumference

The circumference of a circle is not a straight line. But you can measure it by wrapping a tape measure around it. Obviously, the larger the circle, the greater the circumference. But is there a connection between the circumference and the other lengths? What is the connection between the circumference and the diameter?

Exercise 18.4   Ma1

Get several round objects. Measure their diameters and their circumferences. Then divide one by the other, that is, calculate circumference ÷ diameter. Record your results in a table. It might look like this.

object	circumference	diameter	circumference ÷ diameter
cotton reel soup can bicycle wheel 10p coin dinner plate			

What do you notice?

You should find that in all the cases the result of the division is the same, a little bit greater than 3. In fact, for any circle the result of dividing the circumference by the diameter is always the same.

The symbol used for the ratio is π (pi), which is the Greek letter 'p'. It is pronounced 'pie'.

$$\frac{circumference}{diameter} = \pi$$

The number π is very important. It occurs in many areas of mathematics, not only those connected with circles. A scientific calculator has a button which gives you the value of π.

$$\pi = 3.141\ 592\ 654$$

The calculator button usually gives π to an accuracy of nine decimal places. For most uses, we do not need the value so accurately. It is enough to use two or three decimal places, as

$$\pi = 3.14 \ (2\ \text{decimal places}) \quad \text{or}$$
$$\pi = 3.142 \ (3\ \text{decimal places})$$

Exercise 18.5   Ma 1

How can you remember π? One way is by this sentence: the number of letters in the words give the first eight significant figures of π.

How I wish I could recollect pi nicely.
3   1   4   1   5   9   2   6

Try to think up a sentence of your own. Can you go beyond the first eight digits?

> Look back at chapter 7 if you have forgotten about significant figures.

Even the nine decimal places above do not give a 100% accurate value of π. Over thousands of years, mathematicians have tried to find an accurate value for π, or to find approximations to it. In modern times mathematicians found π to an accuracy of hundreds of decimal places. Even greater accuracy is possible with computers, and people have found π to millions of decimal places.

A problem of the ancient Greeks was 'squaring the circle'. This problem is to find a square with the same area as a given circle. For 2000 years mathematicians tried to find a way of doing this, using only a straight edge and compasses. Only in 1882 did a German mathematician called Lindemann find the answer: that you can't do it!

In the next exercise, you look at some of the approximations of pi.

Exercise 18.6

1 In the Bible, I Kings 7, there is a description of a great bowl in front of King Solomon's temple in Jerusalem.

> *And he made a molten sea, ten cubits from the one brim to the other, it was round all about . . . and a line of thirty cubits did compass it round about*

What value does this give for π?

2 In about 225 BC the Greek mathematician Archimedes proved that π lies between $\frac{22}{7}$ and $\frac{223}{71}$. Find these fractions as decimals. Does the value of π given by your calculator lie between them? Which fraction is closer?

3 In about AD 470 a Chinese mathematician called Tsu Ch'ung Chih gave the value $\frac{355}{113}$ for π. What is the difference between this value and the one given by your calculator?

> Use $\pi = 3.142$, and give answers to 3 significant figures.

Whatever the theory, in practice you can use 3.14 or 3.142 for π, or use the π button on your calculator.

Let the diameter of a circle be d and the circumference be C. The definition of π is that

$$\frac{C}{d} = \pi$$

Multiply across by d, and you get a formula giving C in terms of d.

$$C = \pi \times d$$

So, given the diameter of a circle, we can find its circumference.

● *To find the circumference of a circle, multiply the diameter by π.*

Example　A running track is in the shape of a circle with diameter 60 m. What is the length round the track?

The length is the circumference of the circle. To find this, multiply the diameter by π.

$$C = \pi \times d$$
$$= 3.142 \times 60 = 188.5$$

The running track is 188.5 m long.

If we are given the radius instead of the diameter, then just use the fact that the diameter is twice the radius. Let the radius be r:

$$C = \pi \times d$$
$$= \pi \times 2 \times r$$

This is usually written as $2\pi r$.

Example   A cartwheel has radius 26 inches. A strip of iron goes round the edge of the wheel. What is the length of the iron?

The length (in inches) of the iron is the circumference of the circle. Let the circumference be C. To find it, multiply the radius by 2 and by π.

$$C = 2\pi r$$
$$= 2 \times \pi \times 26 = 163.4$$

So the length of the iron is 163.4 inches.

Exercise 18.7

In each of questions 1 to 5 find the circumference of the circle with the given diameter or radius.

1 diameter 10 cm
2 diameter 25 m
3 diameter $5\frac{1}{4}$ inches
4 radius 4 cm
5 radius 17 m
6 The top of a soup can has diameter 6 cm. What is the length round the top?
7 The Moon is about 400 000 km from the Earth. What is the length of its orbit? (Assume the orbit is circular.)
8 The minute hand of a clock is 12 cm long. How far does the tip of the hand travel in 1 hour?
9 The diameter of a computer floppy disk is $3\frac{1}{2}$ inches. What is the circumference of the disk?
10 The radius of the Earth is about 6400 km. What is the length of the equator?

If we know the diameter of a car wheel, then we can find its circumference. On a long journey, the wheel rotates many times.

Example   The wheel of a car has diameter 50 cm. How far does the car travel when the wheel rotates 6000 times?

The circumference of the wheel is $\pi \times 50$ cm, which is 157 cm. If the wheel doesn't skid, then, when the wheel rotates once, the car moves along a distance equal to the circumference of the wheel. So for each rotation, the car moves 157 cm.

So in one rotation the car travels $\pi \times 50$ cm.

s.f. is short for
significant figures

In 6000 rotations the car travels $\pi \times 50 \times 6000$ cm $= 942\,000$ cm (to 3 s.f.)

So the car travels 942 000 cm. This is 9420 m, or 9.42 km.
The car travels 9.42 km.

Exercise 18.8

1 The wheel of a bicycle has diameter 110 cm. How far does the bicycle go when the wheel rotates 100 times?
2 The minute hand of a clock is 8 cm long. How far does the tip of the hand travel during a day?
3 The distance from the Earth to the Sun is about 150 000 000 km. How far does the Earth travel during a period of 40 years? (Assume that the Earth's orbit is circular.)
4 The wheel of a car has radius 30 cm. How far does the car travel when the wheel rotates 100 times?
5 The diameter of a snooker ball is 5 cm. How far does the ball travel when it rolls six times across the table?

Finding the diameter

If you know the diameter of a circle, you can find its circumference. You can also go in the other direction – if you know the circumference of a circle you can find its diameter.
To find the circumference of a circle, multiply the diameter by π.

circumference = diameter $\times \pi$

● *To find the diameter of a circle*, divide *the circumference by π.*

diameter = circumference $\div \pi$

Using letters, if c is the circumference, and d is the diameter

$C = d \times \pi$ and $d = C \div \pi$

Example A circular running track is 400 m long. What is its diameter?

Follow the rule above. Divide 400 by π.

$400 \div \pi = 127$ (to 3 s.f.)

The diameter of the track is 127 m.

Exercise 18.9

In each of questions 1 to 3 the circumference of a circle is given. Find its diameter.

1 4 m

2 23 cm

3 2.55 m

In each of questions 4 to 6 the circumference of a circle is given. Find its radius.

4 200 m

5 39 cm

6 $2\frac{1}{3}$ cm

7 The circumference of a car wheel is 150 cm. What is the diameter of the wheel?

8 A belt of length 21 inches will just go round Anne's waist. Assuming that her waist is circular, what is its diameter?

9 A label of length 32 cm will just go all the way round a soup tin. What is the radius of the tin?

10 The metric system was invented in 1795. The basic unit was the metre, defined so that the circumference of the Earth would be 40 000 000 m. What is the diameter of the Earth?

Area of a circle

What is the area of a circle? This is the Greek problem mentioned on page 327, of 'squaring the circle'. The edge of a circle isn't straight, so we can't cut it up into rectangles or triangles, as we did in chapter 16. We shall have to find the approximate area.

Take a circle of radius 5 cm and cut it into lots of sectors, as shown in the left of the diagram. Then rearrange the sectors head to tail, as in the right side of the diagram.

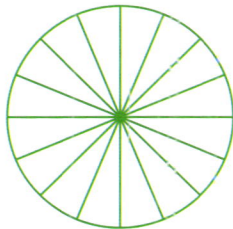

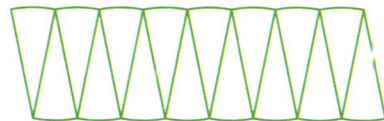

The shape you get is *roughly* a rectangle. So the approximate area of the circle can be found from the area of the rectangle.

height of rectangle = radius of the circle = 5 cm

For the width of the rectangle, notice that all the arcs along the top and bottom of the rectangle make up the circumference of the circle.

> circumference of circle
> $\approx$ top of rectangle + bottom of rectangle

So

> width of rectangle $\approx \frac{1}{2} \times$ circumference of circle
> $= \frac{1}{2} \times \pi \times 10\,\text{cm}$

> The symbol $\approx$ means 'is approximately equal to'.

Put all our facts and formulae together.

> height of rectangle $\approx$ radius of circle $= 5\,\text{cm}$
> width of rectangle $\approx$ half circumference of circle
> $= \frac{1}{2} \times \pi \times 10\,\text{cm}$
> $= \pi \times 5\,\text{cm}$

So

> area of rectangle $=$ width $\times$ height $= \pi \times 5 \times 5\,\text{cm}^2$
> $= \pi \times 5^2\,\text{cm}^2$
> $= \pi \times 25 = 78.5\,\text{cm}^2$
> (to 3 s.f.)

Exercise 18.10

Draw a circle of radius 10 cm, diameter 20 cm. Cut it into 20 equal sectors, and arrange them head to tail. You should get a shape which is roughly a rectangle.

You will need:

- compasses
- scissors
- ruler
- protractor

1 What is the height of your rectangle? (It should be 10 cm.)
2 What is the width of your rectangle? (It should be approximately half the circumference, $\frac{1}{2} \times \pi \times 20$.)
3 What is the area of your rectangle? (It should be $\pi \times 10^2\,\text{cm}^2$.)

Now we'll consider a general circle, of radius r and diameter $2r$.
The height of the rectangle is the radius of the circle, r.
The width of the rectangle is approximately $\frac{1}{2} \times \pi \times 2r$, which is the same as $\pi \times r$.
The area of the rectangle is $\pi \times r \times r$, which is πr^2.

> area $= \pi r^2$

● So the formula for the area of a circle is πr^2. When using it, be careful to square the radius first, and then multiply by π.

Example   The inner edge of a running track is a circle of radius 60 m. What area is enclosed by the track?

The region is a circle of radius 60 m. Apply the formula, squaring 60 and then multiplying by π.

$$\begin{aligned} \text{area} &= \pi \times 60^2 \\ &= 3.142 \times 3600 \\ &= 11\,310 \text{ (to 4 s.f.)} \end{aligned}$$

So the area enclosed is $11\,310\,\text{m}^2$.

Exercise 18.11

In each of questions 1 to 5 find the area of the circle with the radius as given.

1 4 cm
2 40 m
3 12.3 cm
4 $5\frac{1}{4}$ inches
5 0.34 cm
6 The radius of the top of a can is 5 cm. What is the area of the top?
7 The radius of a coin is 2.1 cm. What is the area of the coin?
8 A circle has *diameter* 8 cm. Write down its radius, and hence find the area.
9 Find the area of a circle with diameter 60 m.
10 Measure the diameter of a 2p coin. Find the area of the coin.

SUMMARY

- Many words connected with circles are used in ordinary language.
- The **diameter** of a circle is twice its **radius**.
- The **circumference** of a circle divided by its diameter always has the same value, π.
- To find the circumference of a circle, multiply the diameter by π.
- To find the diameter of a circle, divide the circumference by π.
- The area of a circle of radius r is πr^2. Be sure to square the radius first and then multiply by π.

Exercise 18A

1 In the diagram below, give the mathematical names for the parts of a circle shown in red.

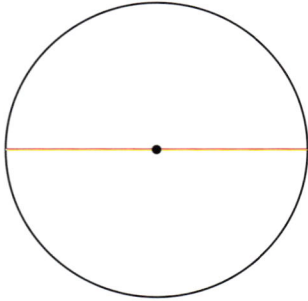

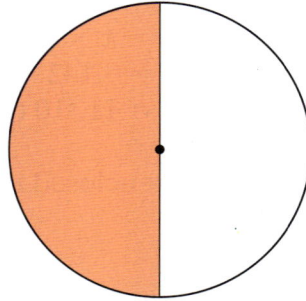

2 The equator is an imaginary circle going round the Earth. Suppose you travel along the equator. What is the mathematical name for the part of the circle you have moved along?

3 A circle has diameter 15 m. What is its circumference?

4 A cotton reel has radius 2 cm. Thread is wound round the reel 300 times. What is the length of the thread?

5 A circle has circumference 33 cm. What is its radius?

6 What is the area of a circle with radius 47 m?

7 What is the area of a circle with diameter 84 cm?

8 A goat is tethered in the middle of a field, on a rope of length 3 m. What area of the field can it graze on?

9 A circle has diameter 4 m. What is the area of a semi-circle taken from the circle?

10 Signals from a radio transmitter can be received for a distance of 100 miles. What area of land can receive the signals?

Exercise 18B

1 Make a copy of the diagram below. Mark on it:
 a the centre
 b the arc AB
 c the chord AB.

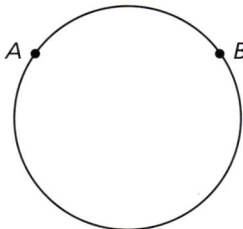

2 A circle has radius 4 cm. What is its circumference?

3 The top of a bottle has diameter 3.2 cm. What is the distance round the top of the bottle?

4 The inner edge of a running track is a circle of radius 78 m. An athlete runs round the track 44 times. How far has she run?

5 The circumference of a wheel is 2.3 m. What is its radius?

6 What is the area of a wheel with radius 6 m?

7 The formula for the area of a circle is πr^2. Find a formula involving π and the diameter d.

8 A semi-circle is taken from a circle of radius 10 cm. What is the total perimeter of the semi-circle?

9 50 m of cotton is wound round a cotton reel of diameter 4 cm. How many times does the cotton go round the reel?

10 When Dido arrived in Libya, the local king, Iarbas, granted her as much land as an ox-hide would cover. The king meant that the hide would be laid flat on the ground, but Dido cleverly cut the hide into a long thin strip. She made the strip into a circle, and claimed all the land within.

Suppose the length of the strip was 100 m.

a What was the diameter of the circle?

b What area of land was enclosed by the hide?

Suppose Dido made the strip into a semi-circle, and used the straight shore line as another boundary.

c How much land would she enclose now?

A version of the story of Dido appears in the *Aeneid*, written by the Roman poet Virgil in about 20 BC. Later she burnt herself to death rather than marry Iarbas, but that is another story with no mathematics in it.

Exercise 18C   Ma1

You will need:
● ruler
● compasses
● scissors

The diameter of a circle is constant. Is the circle the only flat shape for which this is true? Think about this before reading the next paragraph.

Draw an equilateral triangle *ABC*. Put the point of your compasses at *A*, and draw an arc which goes through *B* and *C*. Repeat, with arcs at centre *B* and at *C*. Cut out the shape enclosed by the three arcs.

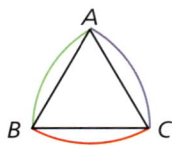

When the shape is placed vertically, how high is the top point above the bottom? Is it always the same?

What other shapes with constant diameter are there? Look in your purse or pocket!

Exercise 18D   Ma1

In exercise 18.4 you found the approximate value of π by measuring circular things. How do mathematicians find π? Obviously they don't do it by measuring soup cans! There are many expressions which get closer and closer to π. The one below was used by a Japanese mathematician called Matsunaga, in 1739.

$$\tfrac{1}{3}\pi = 1 + \frac{1^2}{4 \times 6} + \frac{1^2 \times 3^2}{4 \times 6 \times 8 \times 10} + \frac{1^2 \times 3^2 \times 5^2}{4 \times 6 \times 8 \times 10 \times 12 \times 14} + \ldots$$

The series of fractions goes on for ever, so you will never get $\tfrac{1}{3}\pi$ exactly. Try adding the first three terms, then multiply by 3. How close is this to π? Try again, with the first four terms. Are you getting closer to π? Write down the fifth term, and include that.

19 Standard form

What do you know about astronomy? You probably know that the stars are a vast distance away, so far that a journey to reach them would take longer than a person's lifetime. How far away are they? Two of the brightest stars in the sky are Betelgeuse and Canopus.

Betelgeuse is about 2400000000000000000 metres away
Canopus is about 1000000000000000000 metres away

Both are huge numbers. Which is greater? It isn't easy to tell, just by looking at the numbers. We need a way of writing huge numbers so that we can tell immediately which is greater.

Powers of 10

The numbers above involve very many 0s. For large numbers, each extra 0 at the end represents multiplying by 10. To know immediately how large the number is, we need to know the **powers** of 10.

Hundred	$100 = 10 \times 10 = 10^2$	
Thousand	$1000 = 10 \times 10 \times 10 = 10^3$	
Million	$1\,000\,000 = 10 \times 10 \times 10 \times 10 \times 10 \times 10 = 10^6$	
Billion	$1\,000\,000\,000 = 10 \times 10 \times 10 \times 10 \times 10 \times 10 \times 10 \times 10 \times 10 = 10^9$	

Exercise 19.1

1 Write ten thousand as a product of 10s, and as a power of 10.
2 There are Hindi words *lakh* and *crore*, which mean a hundred thousand and ten million, respectively. Write them out as products of 10s and as powers of 10.
3 There are Japanese number words, *man*, *oku* and *cho*. A *man* is 10000. An *oku* is a *man* × a *man*. A *cho* is a *man* × an *oku*. Write all these as powers of 10.
4 A *googol* is 1 followed by a hundred 0s. Write a googol as a power of 10.
5 In the UK, a *centillion* is 1 followed by 600 zeros. In the USA, it is 1 followed by 303 zeros. Write both of these as powers of 10.

Hundreds and thousands

Numbers aren't always *one* hundred, or *one* thousand. We might need to deal with six hundred, or seven thousand, or nine million. In each case, there is a digit, followed by a string of zeros. We can show this by writing the string of zeros as a power of 10.

$$600 = 6 \times 100 = 6 \times 10^2$$
$$800 = 8 \times 100 = 8 \times 10^2$$
$$7000 = 7 \times 1000 = 7 \times 10^3$$
$$8000 = 8 \times 1000 = 8 \times 10^3$$
$$40\,000 = 4 \times 10\,000 = 4 \times 10^4$$
$$700\,000 = 7 \times 100\,000 = 7 \times 10^5$$

Exercise 19.2

Using the examples above as a guide, copy and complete these lines.

1 $700 = 7 \times 100 =$
2 $400 = 4 \times 100 =$
3 $900 = 9 \times$ _____ =
4 $500 =$ ___ $\times$ _____ =
5 $300 =$ ___ $\times$ _____ =

6 $3000 = 3 \times 1000 =$
7 $5000 = 5 \times 1000 =$
8 $50\,000 = 5 \times 10\,000 =$
9 $90\,000 =$ ___ $\times$ _____ =
10 $600\,000 = 6 \times$ _____ =

Remember:

For the expression 10^8, *the index is 8.*

You can probably see the method. Suppose a number consists of a digit followed by several 0s. Write down the digit, followed by a power of 10. The index of 10 is the same as the number of 0s.

For example, $700\,000$ consists of the single digit 7 followed by five 0s. So multiply 7 by 10 raised to the power 5.

$$700\,000 = 7 \times 10^5$$

single digit → power of 10

● *When a number is written in this way, it is in **standard form**.*

Exercise 19.3

Write the numbers in questions 1 to 6 in standard form.

1 $30\,000$
2 $500\,000$
3 $600\,000$
4 $7\,000\,000$
5 $80\,000\,000$
6 $900\,000\,000$

Look back at exercise 19.1 to see the examples of some Indian and Japanese number words. Write the following in standard form.

7 Seven *lakhs*
8 Nine *crore*
9 Five *man*
10 Eight *cho*

Comparing numbers

Now we are ready to look at two numbers and decide immediately which is the larger. We know that 300 is greater than 70, and that 2000 is greater than 900 because of the number of zeros. Similarly, the most important part of a number in standard form is the power of 10.

- If two numbers have different powers of 10, the one with the higher index is greater.
- If two numbers have the same power of 10, the one with the larger number in front is greater.

For example:

4×10^5 is greater than 7×10^3 (5 is greater than 3)
2×10^8 is greater than 9×10^7 (8 is greater than 7)
7×10^9 is greater than 4×10^9 (both involve 10^9, but 7 is greater than 4)

Exercise 19.4

Identify the larger number in each of the pairs of numbers below.

1 5×10^7 and 8×10^6
2 4×10^7 and 6×10^8
3 2×10^9 and 5×10^7

4 5×10^8 and 4×10^6
5 2×10^5 and 9×10^6
6 6×10^7 and 4×10^7

More than one digit

Of course, not every number consists of a single digit followed by a string of 0s. Consider this number: four million, one hundred and thirty-two thousand.

4 132 000

This number is a bit larger than four million, 4 000 000.

We could write 4 132 000 as 4132×10^3, as there are three 0s at the end. But this does not help us to see immediately that the number is a bit larger than 4 000 000. The most important digit of the number is the 4, the first significant figure. So write the number so that this first significant figure, 4, is immediately in front of a decimal point.

$$4 132 000 = 4.132 \times 1 000 000 = 4.132 \times 10^6$$

Now it is easy to see that the number is about four million.

The **first** significant figure is the **most** significant figure.

To convert 4 132 000 to standard form, we moved a decimal point to the left, until there was only one digit in front of it. We had to move the decimal point six places for this to happen, so we have to multiply by 10^6.

4.1 3 2 0 0 0 (the decimal point moves six times)

● *In general, to put a number in* **standard form***, move the decimal point to the left until there is only one digit in front of it. Count how many times you have moved the point, and that will give you the index of 10.*

Examples Put the number 639 290 in standard form.

Move the decimal point, until it is between the 6 and the 3. You have to move it five times for this, so the index of 10 is 5.

$$639\,290 = 6.3929 \times 10^5$$

Put the number 2539.44 in standard form.

Move the decimal point until only the 2 is in front of it. You have to move the point three times for this to happen.

$$2539.44 = 2.539\,44 \times 10^3$$

Exercise 19.5

Put these numbers in standard form.

1 3521
2 5033.7
3 32 000
4 93 327.9
5 394 833

6 230 000
7 1 219 000
8 23 090 000
9 Two and a half *crores* (see exercise 19.1)
10 Three and a quarter *oku* (see exercise 19.1)

Here are some more comparisons. The method above still holds.

● If two numbers in standard form have different powers of 10, the one with the greater index is the larger.
● If two numbers have the same power of 10, the one with the larger number part is the larger.

For example:

3.2×10^5 is greater than 7.3×10^3 (5 is greater than 3)
2.64×10^8 is greater than 9.1×10^7 (8 is greater than 7)

$$7.325 \times 10^9 \text{ is greater than } 4.983 \times 10^9 \quad (7.325 \text{ is greater than } 4.983)$$
$$3.125 \times 10^3 \text{ is greater than } 3.123 \times 10^8 \quad (3.125 \text{ is greater than } 3.123)$$

Exercise 19.6

For questions 1 to 5 identify the larger number in each pair of numbers.

1 1.644×10^6 and 4.022×10^5
2 2.355×10^7 and 5.14×10^9
3 9.154×10^7 and 5.03×10^7
4 2.32×10^8 and 5.321×10^6
5 5.33×10^9 and 5.12×10^9
6 At the beginning of this chapter the distances to Betelgeuse and Canopus were given. Which star is further away?
7 The population of China is about 1.2×10^9 people, and that of Japan is about 1.3×10^8 people. Which is the larger population?
8 The area of the Atlantic ocean is about $8.22 \times 10^7 \text{km}^2$, and that of the Pacific ocean is about $1.65 \times 10^8 \text{ km}^2$. Which is greater?
9 The diameter of the planet Jupiter is $1.43 \times 10^5 \text{km}$, and that of Saturn is $1.20 \times 10^5 \text{km}$. Which is greater?
10 Below are the populations of five countries. Arrange them in increasing order.

India	Pakistan	Argentina	Indonesia	Germany
8.24×10^8	1.1×10^8	3.2×10^7	1.75×10^8	7.8×10^7

Converting back to ordinary numbers

You can go in the other direction, and convert numbers in standard form to the ordinary way of writing them. This may be necessary if you are talking to someone who hasn't done standard form (or doesn't understand it!).

Suppose you have a large number in standard form. Move the decimal place to the right, as many times as the index of 10. You may have to insert 0s, rather than leaving a blank space.

Take 4.53×10^8. This is a large number, with index 8. Move the decimal point eight places to the right. After the 3 you will have to put in 0s.

Four hundred and fifty-three million

4.5 3 0 0 0 0 0 0

Hence $4.53 \times 10^8 = 453\,000\,000$.

Exercise 19.7

In questions 1 to 5 convert the numbers in standard form to ordinary numbers.

1 3.4×10^6

2 1.7×10^5

3 7.33×10^8

4 4.7354×10^3

5 $7.364\,42 \times 10^9$

6 The population of Nigeria is 1.02×10^8. Write this as an ordinary number.

7 The mass of an ocean liner is 8.3×10^5 tonnes. Write this as an ordinary number.

8 The area of England is about 1.3×10^5 km². Write this as an ordinary number. How do you say this number in ordinary words?

9 The population of the USA is about 2.5×10^8. Write this as an ordinary number. How do you say this number in words?

10 A business tycoon is said to have a fortune of about 3×10^{10} dollars. Write this as an ordinary number. How do you say this sum in words?

Use of a calculator

You can use a scientific calculator to enter numbers in standard form. Usually this is done by a button labelled EXP.

The sequence to enter 4.2×10^5 is

$$\boxed{4} \quad \boxed{.} \quad \boxed{2} \quad \boxed{\text{EXP}} \quad \boxed{5}$$

What you see on the screen will depend on the make of your calculator. It might be one of these.

- Don't press the × button before the EXP button. That may bring up π on the screen!
- Don't enter 10. That may make your answer 10 times too large!

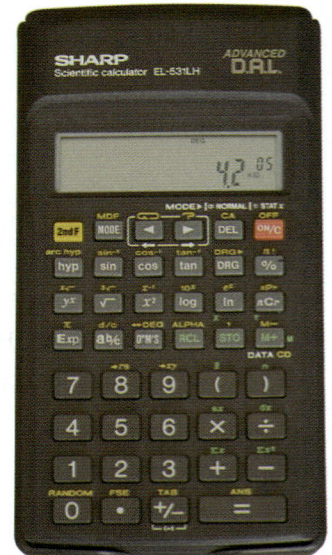

Now press = (or enter). You will see the number converted to an ordinary number, 420 000.

It's a very simple key sequence! Don't make it more complicated.

When you write a number in standard form, write it as 4.2×10^5. Don't write down what your calculator shows – your teacher might have a different calculator!

Exercise 19.8

For questions 1 to 5, enter the number on your calculator. Then press = to convert it to an ordinary number, and write this down.

1 3×10^4
2 5×10^6
3 2.1×10^5
4 8.2×10^6
5 2.36×10^5
6 Try entering 6×10^{125}. What happens?
7 What is the largest number you can enter in your calculator?
8 Long ago, Indian mathematicians were fascinated by very large and very small numbers. Here are the units of an ancient system of measurement of length.

> 7 *primaru* make 1 *renu*
> 7 *renu* make 1 *truti*
> 7 *truti* make 1 *vatayanaraja*
> 7 *vatayanaraja* make 1 *shasharaja*
> 7 *shasharaja* make 1 *edakaraja*
> 7 *edakaraja* make 1 *goraja*
> 7 *goraja* make 1 *eksaraja*
> 7 *eksaraja* make 1 *sarsapa*
> 7 *sarsapa* make 1 *yava*
> 7 *yava* make 1 *anguli parva*
> 7 *anguli parva* make 1 *vatasti*
> 2 *vatasti* make 1 *hasta*
> 4 *hasta* make 1 *dhanu*
> 1000 *dhana* make 1 *krosha*
> 4 *krosha* make 1 *yojana*

a How many *primaru* are there in 1 *yojana*?
b A *yojana* is about $4\frac{1}{2}$ miles. How long is 1 *primaru*?

Arithmetic in standard form

You can multiply and divide numbers in standard form. Here we shall find out what the method is.

Multiplication

Suppose we want to multiply 3×10^5 and 2×10^4. Write both of these as ordinary numbers.

$$300\,000 \text{ and } 20\,000$$

Multiply these together. We get $6\,000\,000\,000$,
   Convert back to standard form: 6×10^9.
   $(3 \times 10^5) \times (2 \times 10^4) = 6 \times 10^9$.

Exercise 19.9

For each of these pairs of numbers in standard form, write them as ordinary numbers, then multiply them together, then convert back to standard form.

1 4×10^5 and 2×10^3

2 3×10^4 and 3×10^5

3 2.3×10^5 and 2×10^3

4 2.1×10^4 and 3×10^4

5 2.5×10^4 and 2×10^5

Have you found the method? Write down what you think the method is. It may help to remember what you found in chapter 13, on indices.
   When multiplying two numbers in standard form, the method is this: multiply the number parts and add the indices.

$$(3 \times 10^8) \times (2 \times 10^9) = (3 \times 2) \times 10^{8+9}$$
$$= 6 \times 10^{17}$$

Now we have this method, we can always use it. We don't have to convert from standard form and back to it.

> **Remember:**
>
> *When multiplying powers, you add the indices.*
> $10^5 \times 10^6 = 10^{11}$

Example A country contains 3×10^7 people, and their average income is £(1.5×10^4). What is the total income of all the people in this country?

We need to evaluate $3 \times 10^7 \times 1.5 \times 10^4$.
   Use the method above, of multiplying the number parts and adding the indices.

$$3 \times 1.5 = 4.5 \qquad 7 + 4 = 11$$

So the total income is £(4.5×10^{11}).

Exercise 19.10

In each of questions 1 to 5, evaluate the product of each pair.

1 3×10^6 and 2×10^7
2 4×10^{14} and 2×10^6
3 3.4×10^9 and 2×10^8
4 1.4×10^{12} and 3×10^9
5 1.5×10^6 and 4×10^{11}
6 A cubic metre of lead has mass 1.2×10^4 kg. What is the mass of 3×10^2 cubic metres of lead?
7 A spaceship travels at 1.5×10^6 metres per second for a period of 3×10^7 seconds. How far does it travel?
8 A firm has 2×10^3 employees, who each earn £(1.8×10^4). What is the total wage bill for these employees?
9 In the Jain religion, there are terms for vast time spans.

> 1 *purvi* = 75 600 000 000 000 years
> 1 *shirsha prahelika* = $(8\ 400\ 000)^{28}$ *purvis*

a Write 1 *purvi* in standard form.
b Write 1 *shirsha prahelika* in standard form.

> According to modern science, the age of the universe is about 1.5×10^{10} years. These lengths of time are vastly greater.

Division

Suppose we want to find 4.2×10^8 divided by 2×10^3.
    Write them as ordinary numbers.

$$420\ 000\ 000 \quad \text{and} \quad 2000$$

Dividing the first by the second, we get 210 000. Converting back to standard form gives 2.1×10^5.

Exercise 19.11

Evaluate these expressions by writing them as ordinary numbers, doing the division, then converting back to standard form.

1 $(6 \times 10^7) \div (2 \times 10^3)$
2 $(8 \times 10^6) \div (2 \times 10^2)$
3 $(9 \times 10^6) \div (3 \times 10^2)$
4 $(4.8 \times 10^7) \div (2 \times 10^4)$
5 $(4.2 \times 10^6) \div (3 \times 10^4)$

Have you found the method? Try to write down the method of dividing a number in standard form by another.

The method is this: divide the number parts, then subtract the indices.

> **Remember:**
>
> *When dividing powers, you subtract the indices.*
> $10^8 \div 10^3 = 10^5$

$$(8 \times 10^{12}) \div (4 \times 10^7) = (8 \div 4) \times 10^{12-7}$$
$$= 2 \times 10^5$$

Example   A star is 5.4×10^{18} metres away. If a spaceship can travel at 3×10^6 metres per second, how long would it take to reach the star?

> **Remember:**
>
> *time = distance ÷ speed*

Divide 5.4×10^{18} by 3×10^6. Divide the number parts and subtract the indices.

$$5.4 \div 3 = 1.8 \quad 18 - 6 = 12$$

So $(5.4 \times 10^{18}) \div (3 \times 10^6) = 1.8 \times 10^{12}$.

The journey would take 1.8×10^{12} seconds.

That's over 50 000 years!

Exercise 19.12

In questions 1 to 5, evaluate the divisions.

1 $(6 \times 10^7) \div (2 \times 10^3)$
2 $(8 \times 10^8) \div (4 \times 10^2)$
3 $(9 \times 10^6) \div (3 \times 10^3)$
4 $(8.4 \times 10^{13}) \div (2 \times 10^6)$
5 $(8.4 \times 10^{23}) \div (7 \times 10^{12})$
6 The speed of light is about 3×10^5 km per second. How long does it take light to travel a distance of 6×10^{43} km?
7 A country has 4×10^7 citizens. Its total wealth is £(8×10^{11}). What is the average wealth, in terms of £ per citizen?
8 A ship is made from metal which weighs 3×10^3 kg per cubic metre. If the ship weighs 4.8×10^7 kg, how many cubic metres of the metal does it contain?

When the answer isn't in standard form

What about $(6.1 \times 10^5) \times (2 \times 10^9)$? When we apply the method, we get

$$(6.1 \times 2) \times 10^{5+9} = 12.2 \times 10^{14}$$

This is the correct number, but it is not in standard form. Can you see why? There should be only one digit before the decimal point, and here there are two. We have to adjust the number so that it is in standard form.

Note that $12.2 = 1.22 \times 10$. This will add another factor of 10 to 10^{14}, changing it to 10^{15}.

$$12.2 \times 10^{14} = 1.22 \times 10 \times 10^{14}$$
$$= 1.22 \times 10^{14+1}$$
$$= 1.22 \times 10^{15}$$

This is very like the 'carry' when we do addition on paper. We 'carry' an extra factor of 10 to 10^{14}.

This is now the correct number, and it is in standard form.
Sometimes the result of a division is not in standard form.

$$(4 \times 10^{10}) \div (8 \times 10^3) = (4 \div 8) \times 10^{10-3}$$
$$= 0.5 \times 10^7$$

This is very like the 'borrow' when we do subtraction on paper. We have 'borrowed' a factor of 10 from 10^7.

Again, this is the correct number, but it is not in standard form. Can you see why? There is no digit in front of the decimal point.
Note that $5 = 0.5 \times 10$. So we need to take one of the factors of 10 from 10^7, leaving 10^6.

$$0.5 \times 10^7 = 0.5 \times 10^{6+1}$$
$$= 0.5 \times 10 \times 10^6$$
$$= 5 \times 10^6$$

This is now correct, and it is in standard form.

Exercise 19.13

Evaluate these expressions, making sure that your answers are in standard form.

1 $(2 \times 10^8) \times (8 \times 10^5)$ **6** $(2 \times 10^9) \div (8 \times 10^3)$
2 $(4 \times 10^6) \times (5 \times 10^8)$ **7** $(3 \times 10^8) \div (6 \times 10^2)$
3 $(7.2 \times 10^9) \times (2 \times 10^4)$ **8** $(1 \times 10^{15}) \div (5 \times 10^7)$
4 $(5 \times 10^5) \times (3 \times 10^{12})$ **9** $(2 \times 10^{23}) \div (5 \times 10^{14})$
5 $(2.5 \times 10^{12}) \times (4 \times 10^3)$ **10** $(7 \times 10^{12}) \div (8 \times 10^5)$

A scientific calculator can do all this arithmetic! Just enter the numbers using the EXP button, and do the arithmetic as normal. The sequence for $(6 \times 10^7) \div (2 \times 10^3)$ is

| 6 | EXP | 7 | ÷ | 2 | EXP | 3 | = |

The answer 30 000 should appear, which is 3×10^4.
Now go through exercise 19.13 again, this time using your calculator to do the arithmetic.

SUMMARY

- We use **standard form** to write large numbers.
- For a number greater than 1, move a decimal point to the left until it is just to the right of the first significant figure. The number of times the point moves gives the index of 10.
- When two numbers are in standard form, it is easy to tell which is the greater.
- You can convert a number in standard form back to an ordinary number. Just move the decimal point to the right the number of times given by the index of 10. You may have to insert 0s.
- When you multiply two numbers in standard form, you multiply the number parts and add the indices.
- When you divide a number in standard form by another, you divide the number parts and subtract the indices.
- After multiplying or dividing numbers in standard form, the result may not be in standard form.

Exercise 19A

1 In the UK, a *quintillion* is 1 followed by 30 zeros. Write this as a power of 10.
2 Write 30 000 in standard form.
3 Write 453 380 in standard form.
4 The number of different selections of numbers in the National Lottery is 13 983 816. Write this fact using standard form.
5 Which is greater, 5.3×10^8 or 8.4×10^7?
6 Write 6.33×10^7 as an ordinary number.

In questions 7 to 10, give your answers in standard form.

7 Evaluate $(2.8 \times 10^4) \times (2 \times 10^7)$.
8 Evaluate $(8.4 \times 10^9) \div (3 \times 10^4)$.
9 Evaluate $(6 \times 10^8) \div (8 \times 10^2)$.
10 One gram of hydrogen contains about 5×10^{27} atoms. How many atoms are there in 3×10^{17} grams of hydrogen?

Exercise 19B

1 Put 543 944 in standard form.
2 Put 382 000 000 in standard form.
3 The Earth's crust was formed 4 500 000 000 years ago. Write this fact using standard form.
4 The number of possible arrangements of *Rubik's Cube* is 43 252 003 274 489 856 000. Write this fact using standard form.
5 Which is greater, 3.65×10^7 or 3.7×10^7?

6 Write $2.643\,844 \times 10^5$ as an ordinary number.

7 Evaluate $(4 \times 10^7) \times (1.5 \times 10^9)$. Give your answer in standard form.

8 How long would it take a swallow, flying at 6×10^4 metres per hour, to travel a distance of 9×10^6 metres? Give your answer in standard form.

9 Evaluate $(8 \times 10^7) \times (3 \times 10^{12})$. Give your answer in standard form.

10 One gram of carbon contains about 4×10^{26} atoms. What is the weight of 3×10^{31} atoms? Give your answer in standard form.

Exercise 19C Ma1

The eternal silence of these immense spaces frightens me. Blaise Pascal (1665)

Many of the examples of this chapter have been about space. The numbers involved in space are so huge that we need standard form to write them down. You could make a class display to show how large the numbers are.

1 Choose a star. Find numerical facts about it, such as its mass, its distance from Earth and its diameter. Put all these numbers in standard form.

2 Suppose a spaceship travels at 4×10^7 metres per second. (This means that it would take 1 second to go round the Earth!) How long would this spaceship take to reach the star?

3 Divide the diameter of your star by the diameter of the Earth. This will show how much wider the star is than Earth. You can show this on a picture.

4 Divide the mass of your star by the mass of the Earth. This shows how much larger the star is than Earth. You could make a comparison, such as 'If the Earth was a pea, this star would be a _____'.

Exercise 19D

The Ancient Greeks did not have a way to write large numbers. The mathematician Archimedes set himself the task of finding the answer to the question 'How many grains of sand would fill the Earth?'. In order to do this, he had to invent his own way of writing large numbers. His method is much more complicated than standard form! With standard form you can answer his question more easily.

The radius of the Earth is about 6400 km. Assume that grains of sand are little cubes of side 0.5 mm. If a sphere has radius r then its volume is $\frac{4}{3}\pi r^3$.

Archimedes, the greatest scientist and mathematician of the ancient world, lived between 287 and 212 BC.
The calculation about the grains of sand is in a short work called *The sand-reckoner*.

20 Volume

You might want to know the volume of things! How much drink is there in a cola can? How much space is there in a room? How much petrol vapour gets sucked into the cylinders of a car? How many scoops of ice cream can you get from a tub?

In Book 1 you worked out the volumes of cuboids. But there are many shapes which are not as simple as this. A drinks can is a cylinder. Some brands of chocolate come in boxes which are prisms or pyramids.

You can find the volume of a cuboid, by filling it with standard cubes of 1 cm by 1 cm by 1 cm. This diagram shows a standard cube, of volume 1 cm^3.

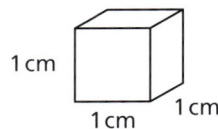

1 cm^3 is read as 'one cubic centimetre', not as 'one centimetre cubed'.

1 cm

1 cm

1 cm

Example Suppose a cuboid is 5 cm long, 3 cm broad and 4 cm high as in the diagram below. Find its volume.

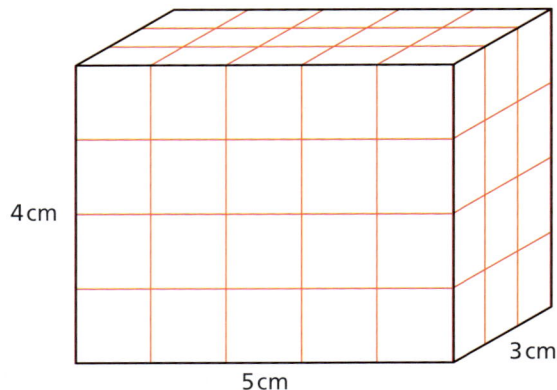

4 cm

5 cm

3 cm

The front of the cuboid is a rectangle 5 cm long and 4 cm high. The front layer of cubes will contain 5 × 4, that is, 20 cubes. There are three layers, so the total number of cubes is 3 × 20, that is, 60.

The volume of the cuboid is 60 cm^3.

We don't have to count the number of standard cubes. We can just multiply together the length, breadth and height.

$$\text{volume of cuboid} = \text{length} \times \text{breadth} \times \text{height}$$

Using letters:

$$V = l \times b \times h$$

Exercise 20.1

(Mainly revision)
Find the volume of each of the cuboids in questions 1 to 4. Lengths are in centimetres.

1

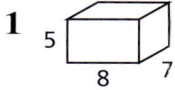

5, 8, 7

2

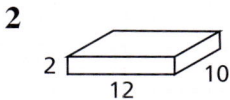

2, 12, 10

3

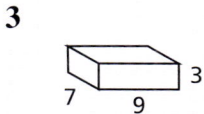

3, 7, 9

4

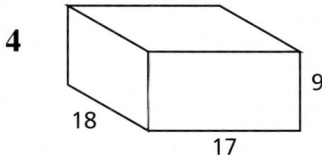

9, 18, 17

5 The interior of a refrigerator is a cuboid, 30 cm by 40 cm by 50 cm. What is its volume?
6 A large room is a cuboid, 4 m high. The floor is a rectangle, 10 m by 15 m. Find the volume of the room.

Fractional volumes

In the example and exercise so far we were able to fill the volumes with standard cubes. That was because all the lengths were whole numbers of the units concerned.

Life isn't always so easy! When we measure a length accurately, we are unlikely to find that it comes to exactly a whole number of centimetres or a whole number of metres. That doesn't matter: we could cut up the standard cubes into pieces, and fill it that way.

In terms of arithmetic, all we do is multiply the length, breadth and height together, and it doesn't matter whether they are whole numbers or fractions. But they must all be in the same units.

Example   A cuboid has length 5 cm, breadth $3\frac{1}{2}$ cm and height 3 cm. Find its volume.

Its volume is found by multiplying these three together.

$$V = lbh$$
$$= 5 \times 3\frac{1}{2} \times 3$$
$$= 52\frac{1}{2}$$

So the volume is $52\frac{1}{2}$ cm³.

Exercise 20.2

Find the volumes of each of the cuboids in questions 1 to 4. Lengths are in metres.

1
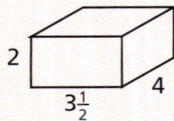

2
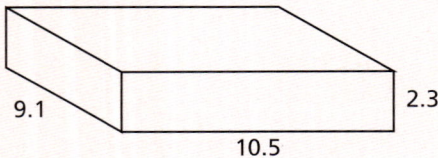

3
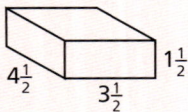

4
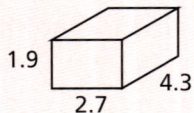

5 A room is a cuboid, with a ceiling $2\frac{1}{2}$ m high. The floor is a rectangle $4\frac{1}{2}$ m by $5\frac{1}{2}$ m. Find the volume of the room.

6 A carton of cereal is a cuboid, which is 4.2 cm by 14.5 cm by 19.6 cm. Find its volume.

Exercise 20.3   Ma1

There are health guidelines to prevent overcrowding. In a classroom, there should be at least 1.8 m² of floor space for each pupil. Measure the floor of your classroom and find its area. How many pupils should it hold? Measure the height of your classroom, and find its volume. What volume of air does each pupil get?

Mixing units

This is something to watch out for. When you are given the dimensions of a cuboid, check to see whether they are all in the same units. If they aren't, then convert so that they are.

Suppose you have a sheet of ice covering a swimming pool. The length and breadth of the pool are probably given in metres, but the depth of ice is probably given in centimetres or millimetres. To find the volume of the ice, convert the lengths so that they are all in the same units.

Example   A swimming pool is 50 m long and 10 m broad, and is covered by a sheet of ice 2 cm deep. Find the volume of the ice.

We could either convert the metres to centimetres, or the centimetres to metres. The numbers will be very large if we take the first choice, so it is easier to take the second.

$$2 \text{ cm} = 0.02 \text{ m}$$
$$V = 50 \times 10 \times 0.02 = 10$$

So the volume of ice is 10 m^3.

If we had converted the metres to centimetres, the volume would be 10 000 000 cm^3.

Exercise 20.4

Find the volume of each of the cuboids in questions 1 to 4. Give your answers in the units suggested.

1 in m^3

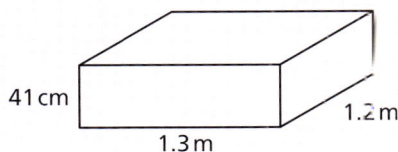

41 cm   1.3 m   1.2 m

2 in cm^3

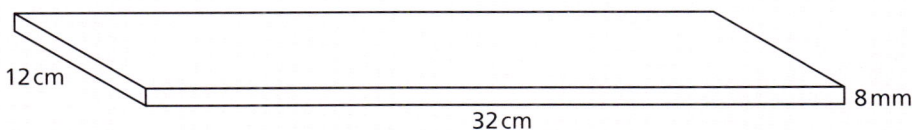

12 cm   32 cm   8 mm

3 in m^3

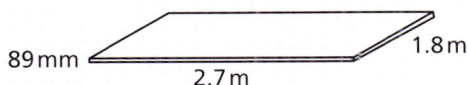

89 mm   2.7 m   1.8 m

4 in cm^3

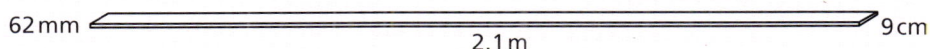

62 mm   2.1 m   9 cm

5 A sheet of hardboard is 1 cm thick. What is the volume of a sheet which is 2 m by 3 m? Give your answer in cubic centimetres.

6 Fine furniture is often covered by a veneer, which consists of thin sheets of wood. Suppose the veneer is 1 mm thick. What is the volume of a rectangular piece of veneer which is 30 cm by 40 cm? Give your answer in cubic centimetres.

7 A road is covered by concrete 8 cm deep. The width of the road is 5 m. What is the volume of concrete in 2 km of the road? Give your answer in cubic metres.

8 Find the volume of concrete on the road of question 7, giving your answer in cubic centimetres and expressing the answer in standard form.

Look again at the cuboid at the beginning of this chapter. Its front face was filled by a layer of 20 standard cubes. There were three of these layers, filling the whole cuboid. Notice that the *area* of the front is 20 cm².

Try for some other cuboids. Take the cuboid shown below, which is 2 cm by 3 cm by 5 cm.

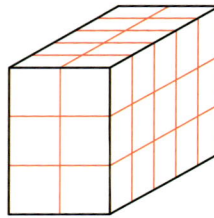

The number of standard cubes needed to fill the front face is 2 × 3, that is, 6. These cubes cover the area of the front face, which is 2 cm × 3 cm, that is, 6 cm². To fill the whole of the cuboid we need 5 layers, so we need 5 × 6, that is, 30 standard cubes. The volume is 30 cm³. The volume can be found by multiplying the front area by the breadth of the base.

Exercise 20.5

Consider a cuboid which is 2 cm long, 3 cm broad and 4 cm high.

1 What is the area of the front face?
2 How many standard cubes do you need to fill the front face?
3 What is the volume of the cuboid?

You should find that, in all cases, the volume of the cuboid is equal to the *area* of the front face, multiplied by the breadth of the base.

The area of the front face gives the number of standard cubes needed for the front layer.

The breadth gives the number of layers to fill the whole cuboid.

volume of cuboid = area of front × breadth

The face doesn't have to be the front face. It could be a side face, or the top or the bottom, and the depth is then the other dimension.

Exercise 20.6

1 A cuboid is 7 cm long, 5 cm broad, 3 cm high.
  a What is the area of the front face?
  b What is the product of the area and the breadth?
  c What is the volume of the cuboid?
2 A cuboid is 8 cm long, 6 cm broad, 2 cm high.
  a What is the area of the front face?
  b What is the product of the area and the breadth?
  c What is the volume of the cuboid?
3 A room is 3 m high. Its floor is a rectangle 4 m by 5 m.
  a What is the area of the floor?
  b What is the product of the area and the height?
  c What is the volume of the room?

Prisms

Rectangular and triangular prisms

The front and the back of a cuboid are the same shape. And if you cut the cuboid parallel to the front, the exposed face (in red) is always the same shape. It is always a rectangle, equal in area to the rectangle on the front.

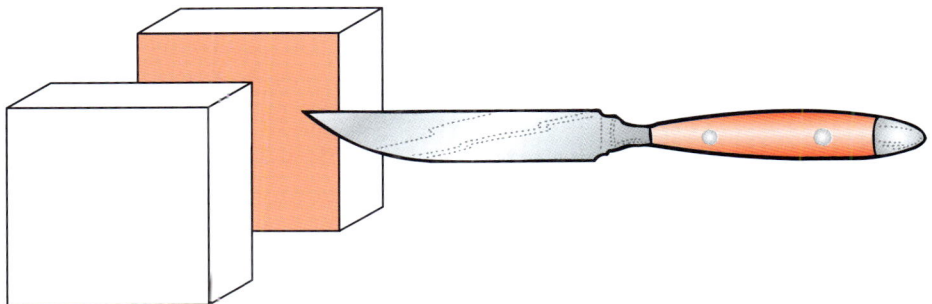

The cut face is a **cross-section**. The cross-section of a cuboid is a constant rectangle. There are many other shapes whose cross-section is constant.

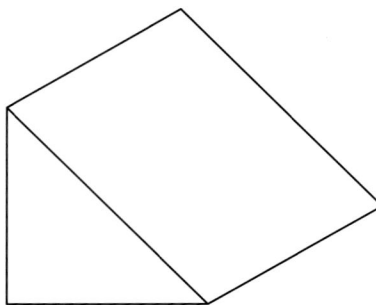

Look at this shape. It is a **triangular prism**. You may have heard of prism binoculars, which use pieces of glass shaped like this.

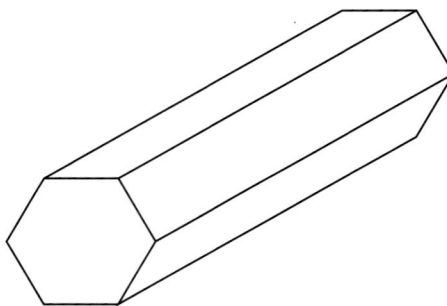

Look at this shape. It is also a prism, with hexagonal cross-section. Some pencils are shaped like this!

How can we find the volume of these shapes? If we try to fill them with unit cubes, they won't fill them completely. There will be jagged edges along the sloping faces which are left unfilled. It won't help us to cut the cubes as we have done before – however finely we chop up the cubes, there will always be a bit of the prism left unfilled.

But how about this! When we look at a cuboid as a prism, we have a formula for its volume.

volume = area of front face × breadth
volume = area of cross-section × breadth

This formula works for all prisms, not just for cuboids.

Let us check the formula for a triangular prism.

Example The prism shown below has a cross-section which is a right-angled triangle, with sides 3 cm, 4 cm and 5 cm. The breadth of the prism is 2 cm. Find its volume.

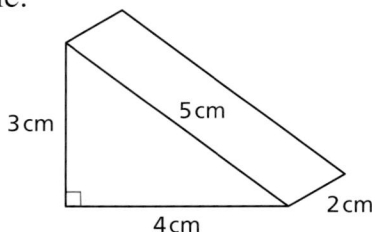

This prism could be made from a 3 cm by 4 cm by 2 cm cuboid by cutting it in half along a diagonal. So the volume of the prism is half the volume of the cuboid.

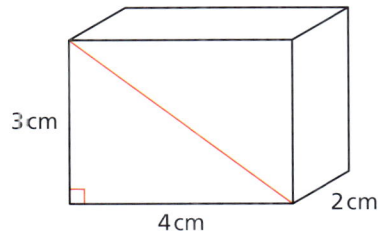

volume of prism $= \frac{1}{2} \times 3\,\text{cm} \times 4\,\text{cm} \times 2\,\text{cm} = 12\,\text{cm}^3$

Now let us find the volume of the prism using the product of the cross-section area and the breadth, as in our formula. The area of cross-section of the prism is given by

area $= \frac{1}{2} \times 3\,\text{cm} \times 4\,\text{cm} = 6\,\text{cm}^2$

The volume of the prism is given by

volume = area of cross-section $\times$ breadth
$= 6\,\text{cm}^2 \times 2\,\text{cm} = 12\,\text{cm}^3$

The result is the same.

Exercise 20.7

The diagram on the right shows a prism whose cross-section is a right-angled triangle with perpendicular sides of length 4 cm and 7 cm. The breadth of the prism is 12 cm.

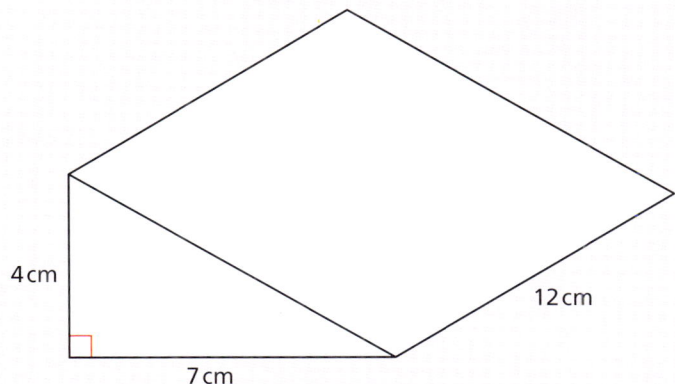

1 Show that the prism could be made by cutting a cuboid in half. Find the volume of the cuboid, and hence find the volume of the prism.
2 Find the area of cross-section of the prism. Multiply this by the breadth of the prism, and hence find the volume of the prism. Are your two answers the same?

In all cases, the volume of a prism is given by this formula.

● *Volume = area of cross-section × breadth*

Exercise 20.8

Find the volume of each of the prisms in questions 1 to 5. Lengths are in centimetres.

1

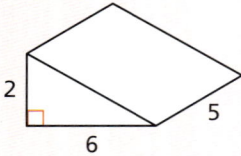

2

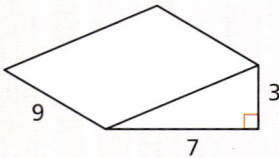

3

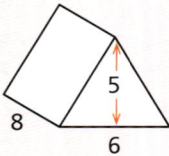

4

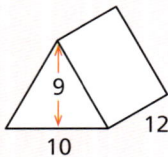

5

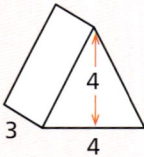

6 The area of the end of a hexagonal pencil is $0.3\,\text{cm}^2$, and the pencil is 12 cm long. What is its volume?

Other prisms

The cross-section of a prism doesn't have to be a simple shape like a triangle or a rectangle. It can, for example, be a trapezium. Often small buildings such as garden sheds are prisms, with a trapezium as cross-section.

Example   This shed has height $2\frac{1}{2}$ m at the front, and 2 m at the back. It is 2 m deep and 3 m wide at the front of the shed. What is its volume?

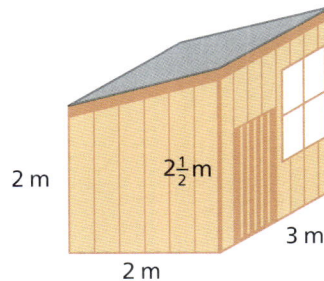

2 m $2\frac{1}{2}$ m   3 m   2 m

Look at a side wall. The two vertical sides are parallel, hence the shape is a trapezium. This is the formula for the area of a trapezium.

Remember the area of a trapezium, from chapter 16.

$$A = \tfrac{1}{2}h(a + b)$$

Put $h = 2$, $a = 2\frac{1}{2}$ and $b = 2$.

In the trapezium formula, h refers to the distance between the parallel sides.

$$A = \tfrac{1}{2} \times 2(2\tfrac{1}{2} + 2)$$
$$= 4\tfrac{1}{2}$$

So the area of the trapezium is $4\frac{1}{2}$ m².
   To find the volume of the shed, we multiply this area by the width of the shed.

$$V = 4\tfrac{1}{2} \times 3$$
$$= 13\tfrac{1}{2}$$

So the volume of the shed is $13\frac{1}{2}$ m³.

Exercise 20.9

Find the volume of each of the prisms in questions 1 to 5. Lengths are in metres.

1

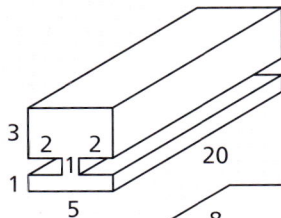

3   2   2   20   1   5

2

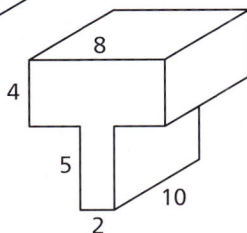

8   4   5   10   2

3

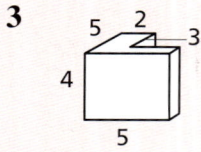

4

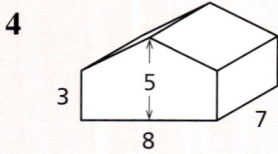

5

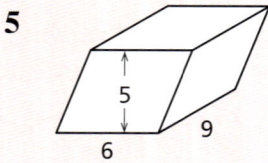

6 Find the area of the side wall of the garden shed shown below. The shed is 3.2 m wide at the front. Find its volume.

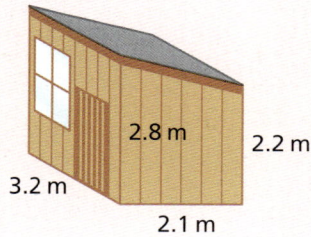

2.8 m
2.2 m
3.2 m
2.1 m

7 The side wall of a house is shown below. Find the area of the wall. If the house is 8 m wide find its volume.

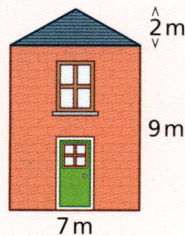

2 m
9 m
7 m

8 A swimming pool is 50 m long and 10 m wide. The shallow end is 1 m deep, and it slopes uniformly to the deep end, which is 4 m deep.
  a Find the area of a side wall.
  b Find the volume of the swimming pool.

Cylinders

Remember, a prism is any shape with a constant cross-section. The cross-section doesn't have to be made from straight lines. The cross-section could be a circle. There is a name for these solids: a **cylinder** is a shape whose cross-section is a circle. The circle is always the same size, wherever you cut the cylinder. So a cylinder is a prism.

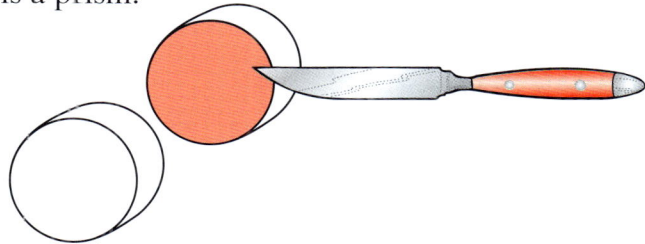

Many things that we buy and use come in cylinders. Look around your home, or as you walk in the street, how many cylinders do you see? There are soup cans, drainpipes, broom handles, the cylinders of a car or motorbike, and so on. Perhaps you can think of many others! There are some objects that you probably don't think of as cylinders – a £1 coin, for example, is a cylinder with a very small height. A length of wire is a cylinder with a very small area of cross-section.

For many of the cylinders above, you might need to know their volume. The volume of a can of soup can tells you how much soup you are getting! The volume of a car's cylinders is a guide for how powerful it is!

A cylinder is a prism, so we can use the same formula for volume.

> volume = area of face × height
> volume = area of circle × height

So, if the cylinder has height h, and the radius of its base is r,

$$V = \pi r^2 \times h = \pi r^2 h$$

> We know how to find the area of a circle (from chapter 18): square the radius, and multiply by π. So, if the radius is r, area of circle = πr^2

Example This soup can has height 12 cm and the radius of its top is 5 cm. What is its volume?

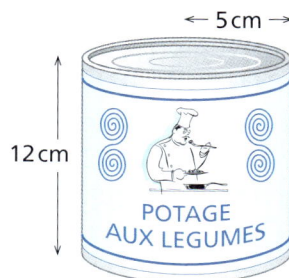

← 5 cm →

12 cm

POTAGE
AUX LEGUMES

Use the formula $V = \pi r^2 h$, substituting $h = 12$ and $r = 5$.

$$V = \pi \times 5^2 \times 12$$
$$= \pi \times 25 \times 12$$
$$= 942 \quad \text{correct to 3 significant figures}$$

So the volume is $942\,\text{cm}^3$.

Exercise 20.10

Find the volume of each of the cylinders in questions 1 to 5. Lengths are in centimetres. Use $\pi = 3.142$ and give your answers correct to three significant figures.

1

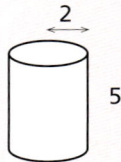

2

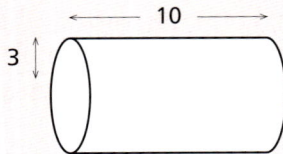

3

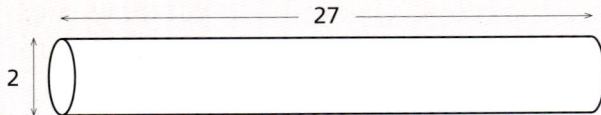

4

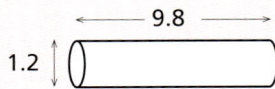

5

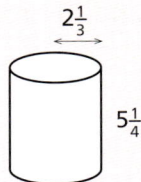

6 Find the volume of a can of beans which has base radius 6 cm and height 10 cm.

7 The base *diameter* of a cylinder is 7 cm, and its height is 20 cm. Find its volume.

8 My car has four cylinders, each of which has height 8 cm and base radius 3.8 cm. What is the capacity of my car's engine?

9 The diameter of a £1 coin is about 2.3 cm and its thickness is about 3 mm. Find the volume of the coin, giving your answer in cubic millimetres.

10 The cross-section of a wire is a circle with diameter 0.8 mm. What is the volume of 10 m of the wire? Give your answer in cubic centimetres.

SUMMARY

- The volume of a cuboid is the product of its length, breadth and height. These don't have to be whole numbers.
- Look carefully at the units in which the measurements are given. If they are different, then convert all the measurements to the same unit.
- The volume of a prism is the product of the area of its cross-section and its height.
- The volume of a cylinder is the product of the area of its base circle and its height.

Exercise 20A

1 A cuboid has width 4 cm, height 3 cm and depth 2 cm. What is its volume?
2 Find the volume of a cube of side 11 cm.
3 A sheet of metal is 0.5 mm thick. What is the volume of a rectangle of the metal which is 88 cm wide and 104 cm long? Give your answer in cubic centimetres.
4 Calculate the volume of a cube of side 0.58 m.
5 The area of cross-section of a prism is 3.4 cm^2 and its length is 124 cm. What is its volume?
6 A box is a cuboid which is 36 cm by 15 cm by 10 cm. The wood is 1 cm thick, so the interior is also a cuboid, which is 34 cm by 13 cm by 8 cm. What is the volume of the wood?
7 The cross-section of a prism is a right-angled triangle with sides 8 cm, 15 cm and 17 cm. The depth of the prism is 12 cm. What is the volume of the prism?
8 The cross-section of a hut is shown below. Lengths are in metres. If the hut is $2\frac{1}{2}$ m wide find its volume.

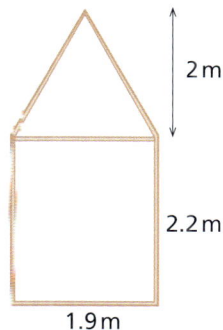

9 A cylinder has height 1.4 m and base radius 0.8 m. What is its volume?
10 A solid tube is 3.8 m long, and its cross-section has radius 2 cm. Find the volume of the tube, giving your answer in cubic centimetres.

Exercise 20B

1 Find the volume of a cuboid with width 3 cm, height 7 cm and depth 4 cm.

2 Find the volume of a cube of side 12.7 m.

3 A film of oil, 0.1 mm thick, floats on top of a reservoir. The reservoir is a rectangle, 800 m by 1200 m. What is the volume of oil? Give your answer in cubic metres.

4 A cube has side 0.45 m. What is the side in centimetres? Calculate the volume of the cube, giving your answer in cubic centimetres.

5 A crystal is a prism with cross-sectional area 9.3 mm². Its length is 62 mm. What is its volume?

6 Find the volume of the prism shown below.

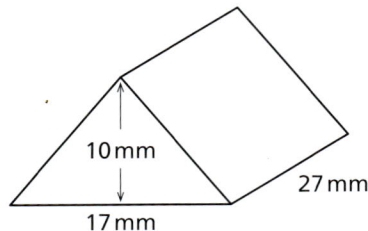

7 The cross-section of a prism is an equilateral triangle, with area 10.5 cm² and length 25 cm. What is the volume of the prism?

8 The cross-section of a house is shown below. Lengths are in metres. If the house is 8 m wide find its volume.

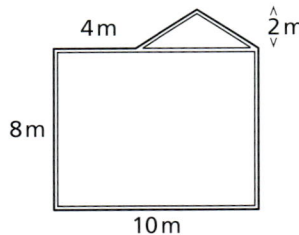

9 A cylinder has height 0.85 m and base radius 3 cm. What is its volume? Give your answer in cubic centimetres.

10 A wire has circular cross-section of diameter 0.6 mm. What is the volume of 3 km of the wire? Give your answer in cubic centimetres.

Exercise 20C   Ma1

Are there any drink cans in your fridge? Do they contain as much liquid as the labels say they do? Measure the height and base radius of a can. Work out the volume – is it what it should be according to the label?

> **Remember:**
>
> *1 ml is the same as 1 cm³.*

Exercise 20D   Ma1

'Eureka!'

In this chapter we have found the volumes of cuboids, prisms and cylinders. All these are regular shapes. How do you find the volume of an irregular shape?

Fill a jar to the brim with water, and put the jar into a bowl. If you immerse an object in the jar, water will spill over into the bowl. The volume of the object is equal to the volume of the water in the bowl, which you can find by pouring it into a measuring jar.

Try this method with an object whose volume you can calculate, such as a cylindrical food can. How accurate is your answer?

Find the volumes of some irregular shapes by this method.

The Greek scientist and mathematician Archimedes needed to find the volume of the king's gold crown, to see whether the king had been cheated by the goldsmith. When climbing into a bath he noticed that water spilled out, and realised that the water's volume was equal to that of his own body. He was so excited that he leapt out of the bath and ran naked down the street, shouting 'Eureka!' (I have found it!)

Use of a calculator

Calculation is quick, easy and accurate with a scientific calculator. Before they became widely available people had to use more difficult methods, involving tables of logarithms, slide-rules, ready reckoners or just paper and pencil! But to get the most out of your calculator you have to become confident and accurate in its use. This section provides tips and practice in the use of a scientific calculator.

What sort of calculator?

Every calculator can work out simple arithmetic, such as 3×15, or $17 \div 4$. A scientific calculator can work out more complicated calculations.

Enter the following into your calculator:

| 4 | + | 3 | × | 5 | = |

With a scientific calculator, you should get the answer 19. If you get any other answer, you may have only an ordinary calculator.

There are many different makes and models of calculator. We hope the instructions here work for yours! If not, read the manual or ask your teacher to find out how to use your calculator.

Calculator exercise 1

Work out the following. Check that you get the correct answer. You don't need to write anything down!

		Answer
1	$5 + 4 \times 3$	17
2	$3 \times 8 + 4 \times 2$	32
3	$6 \times 9 - 5 \times 10$	4
4	$14.5 \times 2 - 3.2 \times 5$	13
5	$4 \times 2.1 - 3 \times 1.7 + 2 \times 0.4$	4.1
6	$2 + 10 \div 2$	7
7	$8 - 7 \div 2$	4.5
8	$15 \div 3 + 16 \div 4$	9
9	$3 \div 4 + 1 \div 8$	0.875
10	$4 \times 2.7 - 3 \div 8$	10.425

Use of brackets

In chapter 5 you learned about brackets. An operation inside brackets is done before the operation outside the brackets. So

$$(4 + 3) \times 5 = 7 \times 5 = 35.$$

A scientific calculator can evaluate this expression directly. Enter the following:

| (| 4 | + | 3 |) | × | 5 | = |

The answer 35 should appear.

Calculator exercise 2

Work out the following. Check that you get the correct answer. You don't need to write anything down!

		Answer
1	$(5 + 4) \times 3$	27
2	$(2.3 + 3.8) \times 1.4$	8.54
3	$(6.8 - 3.9) \times 2.6$	7.54
4	$(0.245 + 0.377) \div 2$	0.311
5	$(37 + 55) \times (0.22 + 0.71)$	85.56
6	$(75.22 + 21.47) \div 0.25$	386.76
7	$23 \div (51 + 29)$	0.2875
8	$3.132 \div (0.24 + 0.63)$	3.6
9	$720 \div (30 \times 80 \times 0.6)$	0.5
10	$(533 + 127) \div (0.22 \times 24)$	125

Memory

A scientific calculator has at least one memory. It is especially useful to store the result of a calculation when you are going to use that result later on! You won't have to key in the number again!

On many calculators, you put a number into the memory with a button labelled Min. You get it out of the memory with a button labelled MR.

If your calculator is different, then read the manual or ask your teacher to find out how to use the memory.

Suppose you have problems about converting several prices in £ to $, at a time when the exchange rate is $1.6435 to the £. Enter this exchange rate into the calculator, then put it into the memory.

| 1 | . | 6 | 4 | 3 | 5 | MIN |

To convert £4 to $, press the following

| 4 | × | MR | = |

The answer 6.574 appears.

To convert £20 to $, press the following

| 2 | 0 | × | MR | = |

The answer 32.87 appears. You see, you didn't have to key in 1.6435 twice!

To convert $50 to £, press the following

| 5 | 0 | ÷ | MR | = |

The answer 30.42, correct to two decimal places, appears.

Calculator exercise 3

1 Press MR to make sure that 1.6435 is still in the memory of your calculator. Use the memory to convert

	Answer
a £26 to $	$42.731
b £44.80 to $	$73.6288
c $131.48 to £.	£80

2 There are 2.54 centimetres in an inch. Put this number in the memory of your calculator, and use it to convert

a 15 inches to centimetres	38.1
b 3.8 inches to centimetres	9.652
c 63.5 cm to inches.	25

3 To convert a price before VAT to a price including VAT, multiply by 1.175. Enter this number into the memory of your calculator. Hence convert these prices before VAT to prices including VAT.

a A computer, costing £780	£916.50
b A holiday, costing £460	£540.50
c A bicycle, costing £162	£190.35

Fractions

Many scientific calculators can deal with fractions. The button for this may be labelled $a^b/_c$. See if you can find it (or something similar) on your calculator.

You can simplify fractions using this button. To simplify $\frac{6}{15}$, enter the following.

| 6 | $a^b/_c$ | 1 | 5 | = |

The answer $\frac{2}{5}$ should appear.

You can do fractional arithmetic using this button. To evaluate $\frac{2}{3} + \frac{1}{4}$, enter the following.

| 2 | $a^b/_c$ | 3 | + | 1 | $a^b/_c$ | 4 | = |

The answer $\frac{11}{12}$ should appear.

You can also enter mixed numbers. To enter $1\frac{3}{4}$, press the following.

| 1 | $a^b/_c$ | 3 | $a^b/_c$ | 4 |

You can now simplify mixed numbers, or do arithmetic on them.

Calculator exercise 4

1 Simplify these fractions using a calculator.

    a $\frac{7}{21}$

    b $\frac{54}{63}$

    c $1\frac{2}{24}$

    d $3\frac{6}{21}$

2 Evaluate these using a calculator. Leave your answers as fractions.

    a $\frac{1}{6} + \frac{1}{5}$

    b $\frac{4}{7} + \frac{3}{8}$

    c $1\frac{2}{3} \times 2\frac{3}{4}$

    d $4\frac{3}{7} - 1\frac{2}{5}$

3 Evaluate these using a calculator. Leave your answers as fractions.

    a $\frac{2}{3} + \frac{3}{4} \times \frac{1}{5}$

    b $\frac{1}{4} \div \frac{1}{3} - \frac{1}{6} \div \frac{1}{5}$

    c $\frac{3}{8} \times \left(\frac{1}{4} + \frac{3}{5}\right)$

    d $\left(\frac{2}{5} - \frac{1}{7}\right) \div \left(\frac{4}{9} + \frac{5}{11}\right)$

Answer

$\frac{1}{3}$

$\frac{6}{7}$

$1\frac{1}{12}$

$3\frac{2}{7}$

$\frac{11}{30}$

$\frac{53}{56}$

$4\frac{7}{12}$

$3\frac{1}{35}$

$\frac{49}{60}$

$-\frac{1}{12}$

$\frac{51}{160}$

$\frac{891}{3115}$

Standard form

In chapter 19 you learned about standard form. Most scientific calculators can enter and use numbers in standard form, often by means of a button labelled EXP. The buttons to enter 5.3×10^7 might be

| 5 | · | 3 | EXP | 7 |

(You don't need to enter 10.)

If the number is not too large, then when you press = it will be converted to an ordinary number. Try this.

| 5 | · | 3 | EXP | 4 | = |

The ordinary number 53 000 should appear.

You can do arithmetic on numbers in standard form. Sometimes the answer will be an ordinary number. To work out $(1.5 \times 10^8) \times (2.4 \times 10^9)$, enter the following

| 1 | · | 5 | EXP | 8 | × | 2 | · | 4 | EXP | 9 | = |

The answer equivalent to 3.6×10^{17} will appear.

Calculator exercise 5

1 Enter these numbers in standard form. What are they in ordinary numbers?

		Answer
a	4×10^3	4000
b	3.1×10^5	310 000
c	9.72×10^6	9 720 000

2 Evaluate these expressions.

a	$(2.7 \times 10^6) \times (4 \times 10^9)$	1.08×10^{16}
b	$(6.3 \times 10^7) \div (1.5 \times 10^3)$	4.2×10^4 (or 42 000)
c	$(4.12 \times 10^{12}) + (3.17 \times 10^{12})$	7.29×10^{12}
d	$(4.27 \times 10^9) - (8.4 \times 10^8)$	3.43×10^9

Spreadsheets

The study of mathematics is being greatly affected by computers! With a computer you can do a vast amount of numerical work in seconds, work which would take you hours or days to do by hand! In this section we present activities on the use of a spreadsheet program.

There are many different spreadsheets available. The activities which follow have been tested on Excel, Lotus and Quattro Pro. We hope they work on other spreadsheets!

Here are three points to remember when using a spreadsheet.

- When you multiply two expressions, you use a $*$ for the multiplication.
- A spreadsheet formula should begin with a $=$.
- You can copy the contents of a cell to another block of cells.

Algebra

In algebra we have expressions like $3x + 7$. The letter x is the variable. If we put $x = 4$, the value of the expression is

$$3 \times 4 + 7 = 19.$$

In a spreadsheet, we enter expressions like =3*A1+7. Here A1 is the variable; it is whatever number is in the cell A1.

In cell A1 enter the number 4. In cell A2 enter the formula =3*A1+7. You should see 19 appear.

Change the number in A1. The value in A2 will also change. Does it always equal $3x + 7$, where x is the number in A1?

Try with different formulae. Some you could try are:

$7x + 8$

$3(x - 4)$ The spreadsheet formula is =3*(A1−4)

$5(x + 3)$

$3(x + y)$ You will need another cell to hold the variable y, perhaps B1.

$x(y + 3)$

$(x + 3)(x - 6)$

Averages

In chapter 15 you worked out averages of sets of data. A spreadsheet can do all this automatically!

Suppose you want to find the mean of these 10 numbers.

32 44 38 29 50 43 27 47 51 48

Enter 32 in A1, 44 in A2, and so on. In A11 enter =SUM(A1..A10)/10. This will give the mean of these numbers.

Try it on some of the other mean problems in chapter 15.

Suppose you want to find the mean of the data from this frequency table.

value	10	20	30	40	50
frequency	3	8	17	13	9

Enter 10 in A1, 20 in A2, and so on. Enter 3 in B1, 8 in B2, and so on.

The C column will hold the products of the values and their frequencies, so in C1 put =A1*B1. You don't have to put four more entries in C2, C3, C4 and C5. Just copy the formula in C1 down to C5.

Finally, in C6 enter =SUM(C1..C5)/=SUM(B1..B5). This will give the mean of the data.

Try it on some of the frequency tables of chapter 15.

Sequences

With a spreadsheet, you can find the terms of a sequence very easily.

Suppose you want the terms of a sequence which starts with 3 and goes up in steps of 7. Enter 3 in A1. In A2 enter the formula =A1+7. Now copy this formula down the A column as far as you want. You will see the terms of the sequence:

3 10 17 24 31 and so on.

Try for some of the other sequences in chapter 12.

Fibonacci sequences

In Book 1 you looked at the *Fibonacci sequence*, in which each term is found by adding the previous two terms.

1 1 2 3 5 8 13 and so on.

Enter 1 in A1. Enter 1 in A2. In A3 enter the formula =A1+A2. Now copy this formula down the A column, as far as you like. You will see the terms of the Fibonacci sequence!

Pascal's triangle

In Book 1 you looked at *Pascal's triangle*, which is the triangle of numbers below. Each term is found by adding the two above it.

```
            1
         1     1
      1     2     1
   1     3     3     1
1     4     6     4     1
```
and so on

Enter 1 in B1. In B2 enter the formula =B1+A2. Copy this formula to the block whose top left corner is B2, and whose bottom right corner is Z26. This block is defined by B2..Z26. You will see the numbers of the triangle appear, though in diagonals rather than in rows.

This is how Pascal himself wrote the numbers, in *Treatise on the arithmetical triangle*. There are much earlier Chinese versions of the numbers, which are in the more familiar triangular form.

Random numbers

In chapter 9 there was an exercise on producing random numbers on a calculator. A spreadsheet also has a facility to generate random numbers between 0 and 1. It may be =RAND or =RAND(). Here we shall use =RAND.

In cell A1 enter =RAND. In cell B1 enter =IF(A1>0.7,1,0). This will be 1 if the random number is greater than 0.7, and 0 if it is less or equal to 0.7.

Copy these values down for 100 rows, that is, to A100 and B100. In cell B101 enter =SUM(B1..B100)/100. This gives the proportion of values greater than 0.7.

What is the experimental probability that the spreadsheet will give a number greater than 0.7?

Dice

You can use a spreadsheet to simulate dice. In A1 enter the formula =INT(6* = RAND+1). This will give a whole number from 1 to 6. Copy the formula down the A column for 60 rows. How many 6s did you get? Is it a fair 'dice'?

Index